AF854856

RAMAKRISHNA AND VIVEKANANDA

New Perspectives

RAMAKRISHNA AND VIVEKANANDA

New Perspectives

Arvind Sharma

STERLING PUBLISHERS PRIVATE LIMITED

STERLING PUBLISHERS PVT. LTD.
L-10, Green Park Extension, New Delhi-110016.
G-2, Cunningham Apartments, Cunningham Road, Bangalore-560052.

Ramakrishna and Vivekananda : New Perspectives
© *1989,* Arvind Sharma

This edition is copyright under the Berne Convention. All rights are reserved. Apart from any fair dealing for the purpose of private study, research, criticism or review as permitted under the Copyright Act, 1956, no part of this publication may be reproduced, stored in a retrieval system, or transmitted, in any form or by any means whatsoever, without the prior permission of the copyright owner. Enquiries should be addressed to Sterling Publishers Pvt. Ltd., L-10, Green Park Extension, New Delhi-110016.

PRINTED IN INDIA

Published by S.K. Ghai, Managing Director, Sterling, Publishers Pvt. Ltd., L-10, Green Park Extension, New Delhi-110016. Printed at Gopsons Papers (P) Ltd., Noida.

for

FAYE SUTHERLAND

Acknowledgements

Permission to publish articles from **Indian Horizons, Journal of Historical Research, Prabuddha Bharat, Religious Traditions, Gandhi Marg, Dialogue, Brahmavadin, Dilip, Studies in Sikhism and Comparative Religion, Journal of the Asiatic Society of Bangladesh** and **The American Academy of Religion** is gratefully acknowledged.

Contents

1

INTRODUCTION

The nineteenth century is regarded by many scholars as a turning point in the history of Hinduism. For "a hundred years ago, at the time of the mutiny, there were many missionaries in India who sincerely believed that within a few generations the whole of India would become Christian."[1] It was felt that the rapid westernization of India was going to leave a "gap which could only be filled by the religion of Christ."[2] It turned out, however, that this gap was filled and is being filled not by Christianity but "by new developments within Hinduism itself."[3]

One such major development, if not the major development within Hinduism, is associated with the figures of Ramakrishna and Vivekananda, who helped to make Hinduism "become a world force".

On account of the immense prestige which Ramakrishna and Vivekananda enjoy in the context of modern Hinduism[4] there has been a tendency to look at them reverently rather than critically and further to look upon them as representing not merely the triumphant vindication of Hinduism but also indirectly encouraging a form of Hindu triumphalism. This book is an attempt to check these tendencies. The main body of the book looks at the life and doctrines of Ramakrishna in particular and that of Vivekananda to a lesser extent in a scholarly and critical spirit, which is here not to be understood as a euphemism for an eruditely negative approach which sometimes characterizes the approach of some modern scholars to Neo-Hinduism. These essays will not therefore please them, and may even displease those who prefer uncritical admiration to analytical presentation of such Hindu heroes as Ramakrishna and Vivekananda. But a Hindu student of Hinduism who sets out to examine his own tradition with a measure of respect as well as detachment must cheerfully accept this situation.

II

In the rest of this Introduction I would like to draw the attention of the reader to what could not be attempted in the body of the book. Just as the nineteenth century produced successful Hindu saints in the form of Ramakrishna and Vivekananda, it also produced despairing Hindu saints such as Pawari Baba and Ramalingar. One may commence with the observation that the nineteenth century has generally been hailed as a period of renaissance in Hinduism.[5] This is corroborated by Dr.S. Radhakrishnan who wrote in 1927: "After a long winter of some centuries, we are today in one of the creative periods of Hinduism."[6] There can be little doubt that Hinduism underwent a transformation during the nineteenth century. It should not be forgotten, however, that this transformation was brought about through the courageous efforts of numerous Hindu reformers who blazed a trail which converted what was the dark night for the soul of Hinduism into one of spiritual enlightenment.

The immensity of the achievement, however, conceals from the eyes of the modern Hindu the difficulty of the task. This can be gauged from the fact that the situation drove even some spiritual stalwarts to despair. Here we shall talk about two of them—Pawari Baba of Ghazipur and Saint Ramalingar of Tamil Nadu.

Pawari Baba of Ghazipur was such a well-known figure in the nineteenth century that he deserves "a place by the side of such men as Dayananda Sarasvati and Ramakrishna."[7] After the passing away of Ramakrishna, Vivekananda even toyed with the idea of becoming his disciple but seems to have been supernaturally prevented from doing so.

What is of importance for the current discussion, however, is the fact that Pawari Baba, it seems, committed suicide out of sheer despair at the condition of India at that time. Max Mueller, writing in 1889, tells us how "his recent death created a painful sensation all over India."[8] He goes on to say:

> He had lived for about thirty years at Ghazipur, and was venerated as a Saint by the whole native community. The last nine years, however, he had almost entirely withdrawn from the world, living by himself in a compound surrounded by high walls and protected by a formidable gate. Inside there was a small flower-garden, a well, a small temple, and his own dwelling-house, which con-

> sisted of one room. He never allowed the gate to be opened, and no one ever saw him except his younger brother. Once every week or ten days, however, he would come up to the gate and converse for a few moments from within with any one who happened to be there. His younger brother always remained within calling-distance. But though his saintly brother had told him that he could not any longer bear the misery which the Kali-yuga, i.e. the present age, had lived for many years cannot be doubted, nor the discovery of The venerable man, after taking his usual bath and performing his devotions, seems to have covered his whole body with clarified butter, to have sprinkled himself all over with incense, then to have set fire to the four corners of his lonely house, and when the flames had taken hold of it on all sides, to have deliberately thrown himself into the fire, thus performing his last sacrifice. Before anybody could rescue him, the old man was burnt to ashes, and what remained of him was consigned with due ceremony to the sacred waters of the Ganges. All this happened only a few months ago. It is always difficult to get an exact account of anything that happens in India. The conflagration of the house in which the old Saint had lived for many years cannot be doubted, nor the discovery of his burnt body. But some of his friends, unwilling to admit his self-immolation, ascribe the fire to an accident, while others consider his voluntary sacrifice as the proper ending to his saintly life.[9]

Saint Ramalingar (1823-1874) presents another interesting case of how even spiritual stalwarts became filled with despair at the condition of the people. As V.A. Devasenapathy points out, he "belongs to the long line of great saints who have adorned our country from time immemorial. He lived in the nineteenth century A.D. and was thus a contemporary of Sri Ramakrishna"[10] though he is little known outside Tamil Nadu.

Saint Ramalingar is believed to have disappeared divinely as we shall see later. We learn that he resumed his lecturing campaign in the last days of January 1874, when he expressed bitter sorrow and disappointment at the reluctant behaviour of the people who thronged around him. To quote the words of his chief disciple, Thozhuvur Velayuda Mudaliar, Ramalingar repeatedly exclaimed:

> You are not fit to become members of this Society of Universal Brotherhood. The real members of that Brotherhood are living far

> away, towards the North of India. You do not follow the principles of my teachings. You seem to be determined not to be convinced by me. Yet the time is not far off when persons from Russia and America (these two countries were always named) and other foreign lands will come to India and preach to you this same doctrine of Universal Brotherhood. Then only you will know and appreciate the grand truths that I am vainly trying to make you accept. You will soon find that the brothers who live in the far north will work a great many wonders in India and thus bring incalculable benefits upon this, our country.[11]

On the 30th of January, 1874, he locked himself up in a room and disappeared. His parting words:

> O, My brothers, I was keeping my shop open so far under the behest of God, Arul - Perum - Jothi; none of you seemed to benefit by it; I have now decided to close this; I will not be visible to your eyes for a specific period, although I will be universally present in the world; I will reappear again at the proper time; till then, take to the path of Ahimsa, worship God in the form of this light and attain salvation.[12]

The two cases are not on all fours, especially as Saint Ramalingar appears more optimistic, even prophetic, when compared with Pawari Baba. Nevertheless the note of pessimism and despair is detectable in the case of both, even though Pawari Baba seems to have been more reclusive of the two, Saint Ramalingar being socially more active. If anything, this should serve to remind us of the heavy odds which the regenerative forces of Hinduism managed to overcome, odds which reduced some of its best products to despair.

III

The misgivings and pessimism of the despairing saints notwithstanding, Hinduism held its ground in the nineteenth century, and "at the very moment when it seemed most likely to have been overwhelmed by the assault of the West."[13] We now turn to a study of two figures who perhaps did as much if not more than anyone else to ensure such an outcome: Ramakrishna and Vivekananda.

References

1. A.L.Basham, "Hinduism" in R.C. Zaehner, ed., *The Concise Encyclopedia of Living Faiths* (Boston: Beacon Press, 1959) p.259.
2. Ibid.
3. Ibid.
4. See D.S. Sarma, *Studies in the Renaissance of Hinduism in the Nineteenth and Twentieth Centuries* (Benaras Hindu University, 1944) Chapters VI, VII.
5. See R.C. Majumdar, H.C. Raychaudhuri and Kalikinkar Datta, *An Advanced History of India* (London: Macmillan & Co., 1950), pp.876-887; D.S. Sarma, *Hinduism Through the Ages* (Bombay: Bharatiya Vidya Bhavan, 1973) Part III, etc.
6. S. Radhakrishnan, *The Hindu View of Life* (London: Unwin Books,1965) p.92.
7. F. Max Mueller, *Ramakrishna: His Life and Sayings* (London: Longmans, Green & Co.,1898) p.15.
8. "His name, Pawari, sometimes spelt Pahari, is explained as a contraction of Pavanahari, he who lives on air"(ibid.).
9. Ibid., pp.14-15.
10. P.Mutharasu, *The Life of Saint Ramalingar*(Tirunelveli:Samarasa Sanmarga Sangam, 1961),Introduction.
11. Ibid., pp.106-107.
12. Ibid., p.107. The following extract from the South Arcot District Manual by Mr J.H.Garstin, ICS, collector of South Arcot District, may be of some interest:

    "PARVATIPURAM: A village of 1189 inhabitants lying 23 miles south west of Cuddalore on the Vriddha-chalam road. The place is connected with one Ramalinga Paradesi, a somewhat curious example of a latter day saint who has been almost deified by his followers.

    Born in 1823 in the Chidambaram Taluk of Vellala parents in humble circumstances, he developed while still little more than a boy, an undeniable talent for versification, and his poems brought him into notice. They dealt with religious matters; some of them, like those of the famous Saivite saints of old, were composed in eulogy of the merits of the deities of certain shrines such as the temples at Tiruttani Tiruttani in north Arcot and Tiruvottiyur near Madras; others took for their subject the beauties of the higher life. It was these that led to his becoming gradually regarded as a spiritual guide and teacher. After visiting many of the well-known sacred places of the south, he finally settled at Karunguli, the next village to Parvatipuram. At its height, his influence must have been very real, his admirers and disciples, who included even level-headed Government officials, are said to have changed their residences and gone to live where they could be constantly near him. About 1872, the curious octogon-shaped *Sabha* with the domed roof which is to be seen at Vadalur, a hamlet of Parvatipuram, was erected by him from subscriptions. It is said that the spot was chosen because from it are visible the four great towers of Nataraja's shrine at Chidambaram. It isn't an ordinary temple, the detail of the worship in it being unusual. Ramalinga Paradesi seems to have persuaded his disciples that they would rise again from the dead and he consequently urged that burial was preferable to cremation. Even Brahmins are said to have been buried in this belief and people who died in other villages were in several cases brought to Vadalur and interred there. In 1874, he locked himself in a room (still in existence) in Mettukuppam (hamlet of Karunguli), which he used for *samadhi* or mystic meditation, and instructed his disciples not to open it for sometime. He has never been seen since, and the room is still locked. It is held by those who still believe in him that he was miraculously made one with his god and that in the fulness of time he will reappear to the faithful.

Whatever may be thought of his claims to be a religious leader, it is generally admitted by those who are judge of such matters, that his poems, many of which have been published, stand on a high plane, and his story is worth noting as an indication of the directions which religious fervour may still take" (ibid., pp.117-118).

13. Ainslie T. Embree, ed., *The Hindu Tradition* (New York: Vintage Books, 1972) p.304.

2

IS MYSTICAL EXPERIENCE UNITIVE OR DIVERSE: THE EVIDENCE FROM RAMAKRISHNA

A considerable debate seems to have developed in recent times around the question: is mystical[1] experience unitive or diverse? The purpose of this paper is to participate in that debate from a Hindu point of view, as represented by experiences of Ramakrishna.

I

One may begin by clarifying the contours of the debate first. According to one school, the mystical experience is essentially unitive. This unitive nature of the mystical experience may be understood in different ways by members of this school, but the emphasis comes to rest on the unitive nature of the mystical experience. Thus William James writes:

> This overcoming of all the usual barriers between the individual and the Absolute is the great mystic achievement. In mystic states we both become one with the Absolute and we become aware of our oneness. This is the everlasting and triumphant mystical tradition, hardly altered by difference of clime or creed. In Hinduism, in Neoplatonism, in Sufism, in Christian mysticism, in Whitmanism, we find the same recurring note, so that there is about mystical utterances an eternal unanimity which ought to make a critic stop and think, and which brings it about that the mystical classics have, as has been said, neither birthday nor native land. Perpetually

> telling of the unity of man with God, their speech antedates languages, and they do not grow old.[2]

The same message is delivered by S. Radhakrishnan with an air of messianic optimism when he writes:

> . . . it seems to me that in the mystic traditions of the different religions we have a remarkable unity of spirit. Whatever religions they may profess, the mystics are spiritual kinsmen. While the different religions in their historical forms bind us to limited groups and militate against the development of loyalty to the world community, the mystics have always stood for the fellowship of humanity. They transcend the tyranny of names and the rivalry of creeds as well as the conflict of races and the strife of nations. As the religion of spirit, mysticism avoids the two extremes of dogmatic affirmation and dogmatic denial. All signs indicate that it is likely to be the religion of the future.[3]

Another scholar who belongs to the same school is Frithjof Schuon. While William James sees the roots of the unitive nature of mysticism in its perennial aspect and S. Radhakrishnan in the universal aspect, Frithjof Schuon tends to see it in the translogical aspect of the situation.[4] He sees in gnosis "the transcendent unity of religions", the last expression serving as the title of one of his books. It may be noted that these scholars perceive the unity among the religions of the world not at the morphological[5] or the ethical[6] level but at the mystical.[7]

II

A considerable body of literature challenges the statement that the mystical experience is unitive. The basic arguments adduced in this body of literature may be summarized as follows:

(1) Some scholars make the point that

. . . there is no such thing as mysticism in the abstract, that is to say, a phenomenon or experience which has no particular relation to other religious phenomena. There is no mysticism as such, there is only the mysticism of a particular religious system, Christian, Islamic, Jewish and so on.[8]

More dramatically phrased, the question is: why is it that Muslim mystics do not have visions of Hindu deities or Christian mystics of Muhammad if mysticism represents a unitive common ground among religions?[9]

(2) Other scholars have argued that the mystical experience is not merely historically but also essentially differentiated. Thus R.C. Zaehner "tried to establish different kinds of types of mysticism: of isolation, of separation of spirit and matter, eternity from time; pantheistic or 'para-enhenic' in which the soul is the universe—all creaturely existence is one; the theistic in which the soul feels identified with God; and the beatitude, with its hope of deification, when the 'perishable puts on the imperishable'."[10]

R.C. Zaehner insists on the distinction between sacred and profane mysticism[11] and tries to explain alleged parallels, say between Hindu and Islamic mysticism, in terms of the influence of one on the other.[12] However, he also finds parallels across traditions for his view that the mystical tradition within the different traditions itself is internally differentiated.[13] The main element which he identifies here is "the great divide ... between those types of mysticism which regarded love as being the central phenomenon of the whole experience and those which, disregarding love altogether, expressed themselves exclusively in terms either of unity or of the escape or 'liberation' from time, the phenomenal world, and all that conditions it."[14] Thus of all the distinctions drawn within 'genuine' religious mysticism this one between love-mysticism and knowledge-mysticism occupies a central place for R.C. Zaehner and the scholars who follow him."[15] Here, then, are two distinct and mutually opposed types of mysticism—the monistic and the theistic",[16] to which may be added a third category of nature mysticism.[17]

Thus it is basically on these two grounds—that mysticism is both historically and internally differentiated—that the claim of its unitive nature has been under attack[18] in recent times.[19]

III

The purpose of the rest of this essay is to examine these criticisms of the unitive nature of the mystical experience in the light of the experiences of Ramakrishna Paramahamsa. The reasons for doing so are that to test these criticisms about the unitive nature of mysticism, we should perhaps examine the actual experiences of a mystic and that, fortunately, in the case of Ramakrishna (1836-1886 A.D.) we possess remarkably detailed and trustworthy accounts of his experiences.[20]

A survey of the religious experiences of Ramakrishna reveals that although he was a Hindu, he had mystical visions of Christ and Muhammad. Once he saw pictures of Jesus in a parlour, one of which was of child Jesus in his mother's lap. His biographer records:

> The Master used to say that he sat one day in that parlour and was looking intently on that picture and thinking of the extraordinary life of Jesus, when he felt as if the picture came to life and effulgent rays of light, coming out from the bodies of the mother and the child, entered into his heart. . . .[21]
>
> The biographer also records that on the third day of his devotion to Jesus . . . the Master saw, while walking under the *Panchavati,* that a marvellous god-man of very fair complexion was coming towards him, looking steadfastly at him. As soon as the Master saw that person, he knew that he was a foreigner. He saw that his long eyes had produced a wonderful beauty on his face, and the tip of his nose, though a little flat, did not at all impair the beauty. . . Very soon the person approached him and from the Master's pure heart came out the words with ringing sound, "Jesus, Jesus the Christian, the great *Yogi,* the loving son of God and one with the Father, who gave his heart's blood and put up with endless torture in order to deliver men from sorrow and misery!"[22]

Similarly, Ramakrishna also had a vision of the Prophet. He was initiated into Islamic practices by a Hindu convert, Govind Rai, and

> . . . engaged himself in practising Islam according to its prescribed rules. The Master said, "I then devotionally repeated the holy syllable 'Allah', wore cloth like the Muslims, said Namaz thrice daily and felt disinclined even to see Hindu deities not to speak of saluting them, inasmuch as the Hindu mode of thought vanished altogether from my mind. I spent three days in that mood, and had the full realization of the result of the practices according to that faith". At the time of practising Islam the Master at first had the vision of an effulgent, impressive personage with a long beard. . .[23]

Before Ramakrishna turned to experimenting with non-Hindu religions, he underwent several phases of Hindu religious practices. From the point of view of this essay, two experiences of his are of capital importance. His first major mystical experience was a vision of *Kali*—the goddess. He describes it in these words.

> I was then suffering from excruciating pain because I had not been blessed with a vision of the Mother. I felt as if my heart were being squeezed like a wet towel. I was overpowered by a great restlessness, and a fear that it might not be my lot to realize Her in this life. I could not bear the separation any longer; life did not seem worth living. Suddenly my eyes fell on the sword that was kept in the Mother's temple. Determined to put an end to my life, I jumped up like a madman and seized it, when suddenly the blessed Mother revealed herself to me, and I fell unconscious on the floor. What happened after that externally, or how that day or the next passed, I do not know but within me there was a steady flow of undiluted bliss altogether new, and I felt the presence of the Divine Mother.[24]

This experience obviously corresponds to the theistic type of mystical experience. Ramakrishna had several such experiences. Towards the end of 1865, however, he was coached in the monistic approach to reality and Ramakrishna again describes his experiences in his own words thus:

> After the initiation, my *guru* began to teach me the various conclusions of the *Advaita Vedanta* and asked me to withdraw the mind completely from all objects and dive into the *Atman* (self). But in spite of all my attempts, I could not cross the realm of name and form, and bring my mind to the unconditioned state. I had no difficulty in withdrawing the mind from all objects except one, the all too familiar form of the Blissful Mother—radiant and of the essence of Pure Consciousness—which appeared before me as a living reality, preventing me from passing beyond the realm of name and form. Again and again I tried to concentrate my mind upon the teachings of *Advaita* (non-dualism), but every time the Mother's form stood in my way. In despair I said to the guru, "It is hopeless. I cannot raise my mind to the unconditioned state and come face to face with the *Atman* (Self)". He grew excited and sharply said, "What! You can't do it! But you have to". He cast his

> eyes round, and finding a piece of glass, took it up; then, pressing the point between my eyebrows, he said, "Concentrate the mind on this point!" Then with a stern determination, I again sat to meditate and as soon as the gracious form of the Divine Mother appeared before me, I used my power of discrimination as a sword and with it severed her form in two. There remained no more obstruction to my mind, which at once soared beyond the relative plane, and I was lost in *samadhi* (subconsciousness)![25]

Thereafter Ramakrishna returned to normal consciousness only after great effort. In the description which follows, one sees the emergence of the theistic element after the monistic experience:

> After his *guru* had left Dakshineswar, Sri Ramakrishna was determined to remain in a state of absolute identity with *Brahman*. He accordingly meditated and entered into *nirvikalpa samadhi* again. Referring to this period of his life, he used to say, "For six months at a stretch I remained in that state whence ordinary men can never return—the body falling off after twenty-one days like a seared leaf. I was not conscious of day and night; flies would enter my mouth and nostrils just as they do in a dead body, but I did not feel them; the hair became matted with accretions of dust. There was no chance for the body to survive, and it would certainly have perished but for the kind ministrations of a monk who was present at Dakshineswar at the time. He realized the state of my mind and also understood that this body must be kept alive at any cost, since it was to be of immense good to the world. He therefore busied himself in preserving this body. He would bring food regularly to me and try various ways to bring my mind down to the consciousness of the relative world, even by beating me with a stick. As soon as he found me to be a little conscious, he would press some food into my mouth, only a bit of which reached the stomach; and there were days in which all his efforts would be in vain. Six months passed in this way. At last I received the Mother's command, 'Remain on the threshold of relative consciousness *(bhavamukha)* for the sake of humanity'. Then I was laid up with a terrible attack of dysentery; an excruciating pain in the stomach tortured me day and night; it went on for six months. Thus only did the mind gradually come down to a lower level and to consciousness of the body. I became a normal man; but before that, at the slightest opportunity, the mind

> would take a transcendental flight and merge in the *nirvikalpa samadhi* . . ."
> Sri Ramakrishna . . . had thus gone through the whole range of Hindu disciplines in the course of twelve years, from his eighteenth year to his thirtieth. He had become a Paramahamsa.[26]

A survey of the various religious experiences of Ramakrishna also reveals the fact that he himself distinguished between mystical states of various kinds, which are called *samadhis*. Thus he asks, "What is *samadhi?*" and answers.

> It is the complete merging of the mind in God-consciousness. The *jnani* experiences *jada samadhi* in which no trace of "I" is left. The *samadhi* attained through the path of *bhakti* is called *chetana samadhi*. In this *samadhi* there remains the consciousness of "I" — the "I" of the servant-and-Master relationship, of the lover-and-Beloved relationship, of the enjoyer-and-Food relationship. God is the Master; the devotee is the servant. God is the Food, and the devotee is the enjoyer. "I don't want to be sugar. I want to eat it."[27]

Out of the various descriptions of the various types of *samadhis* by Ramakrishna,[28] this description has been chosen because of its correspondence to the basic divide in mystical experience developed by R.C. Zaehner, who contrasts devotional mysticism with gnostic mysticism.[29] Yet, though for R.C. Zaehner this means the fragmentation of mysticism—not so for Ramakrishna. Ramakrishna, though he describes various kinds of *samadhis*, also insists that they tie into the same Reality. His position is best stated in his own words:

> Do you know what I mean? *Satchidananda* is like an infinite ocean. Intense cold freezes the water into ice, which floats on the ocean in blocks of various forms. Likewise, through the cooling influence of *bhakti*, one sees forms of God in the Ocean of the Absolute. These forms are meant for the *bhaktas*, the lovers of God. But when the sun of knowledge rises, the ice melts; it becomes the same water it was before. Water above, the water below, everywhere nothing but water. . .
> But you may say that for certain devotees God assumes eternal forms. They are placed in the ocean where the ice doesn't melt at all. It assumes the form of quartz.[30]

Moreover, Ramakrishna's life offers graphic illustrations of what is called Nature Mysticism as well. At these moments, his sense of identity with the rest of the universe was so complete that a thousand "instances of his hypersensitiveness might be related".

> A blow given to a man in the street by a furious enemy, left its mark on the flesh of Ramakrishna. His nephew saw his back red and inflamed at the sight of a man whose back was scored with the whip. And Girish Chandra Ghosh, whose witness is unimpeachable, has certified to the fact of his stigmata. This spiritual contact with all forms of life made him at one even with animals and plants. It has been said of him that he felt a brutal step upon the earth as it were upon his own heart.[31]

IV

What are we, then, to make of the view that mysticism is heterogeneous in the light of the experiences of Ramakrishna. It is quite obvious that his case contradicts both the historical and the ontological argument which challenges the unitive nature of mysticism. Ramakrishna did have visions of deities other than Hindu,[32] and he did have experiences of the various types of mystical states—especially the monistic and theistic between which so many scholars like to distinguish. In this connection, two important points should not be overlooked.

One is that these experiences are not to be called unitive because they were unified in the person of Ramakrishna. The unitive nature of mysticism is a comment not about the experiencer but what is experienced.

The other point to remember—and this appears to be the key point—is that Ramakrishna was *not* claiming that all his experiences of the Reality were the same. [By extension, it is not the Hindu mystical claim that mystics have the same experience of Reality]. It was Ramakrishna's claim, not that mystics have or he had the same experience of Reality, rather that it is the same Reality which was experienced by him, time and again or is experienced by the mystics in diverse ways. Ramakrishna elaborated this point in a large number of sayings:

> God has made different religions to suit different aspirants, times and countries. All doctrines are only so many paths; but a path is by

> no means God Himself. Indeed, one can reach God if one follows any of the paths with whole hearted devotion. One may eat a cake with icing either straight or sidewise. It will taste sweet either way.
> As one and the same material, water, is called by different names by different peoples, one calling it water, another eau, a third aqua, and another pani, so the one Everlasting-Intelligent-Bliss is invoked by some as God, by some as Allah, by some as Jehovah, and by others as Brahman.
> As one can ascend to the top of a house by means of a ladder or a bamboo or a staircase or a rope, so diverse are the ways and means to approach God, and every religion in the world shows one of these ways.
> As the young wife in a family shows her love and respect to her father-in-law, mother-in-law, and every other member of the family, and at the same time loves her husband more than these; similarly, being firm in thy devotion to the deity of thy own choice, do not despise other deities, but honour them all.
> Bow down and worship where others kneel, for where so many have been paying the tribute of adoration, the kind Lord must manifest himself, for he is all mercy.
> The devotee who has seen God in one aspect only, knows him in that aspect alone. But he who has seen him in manifold aspects is alone in a position to say, "All these forms are of one God and God is multiform." He is formless and with form, and many are his forms which no one knows.[33]

It will be noticed that in the second saying it is *not* claimed that the names for water are not different; it is rather water which is not different. Similarly in the third saying it is *not* claimed that ladder, bamboo, staircase, and rope are the same; but they have a commonality of being means of approach. More pointedly, in the last saying—it is not claimed that God is uniform—it is clearly stated that God is multi-form but it is the *same* God who assumes different forms. Similarly, elsewhere, it is *not* claimed that there is no difference between Krishna and Christ—what is claimed is that it is the same Saviour who rises up in different forms.[34]

It is difficult to avoid the conclusion that the scholars who have gone about classifying types of mystical experience, have overlooked this central claim in much of Hindu mysticism, that it is not the experiences which are to be taken as identical, but the Reality experienced through them. Or perhaps the point one is trying to make was made centuries ago—and more pointedly, in Persia by Rumi when he pointed out the roads are different and diverse, *not* the goal.

> Though the roads are various, the goal is one. Do you not see that there are many roads to Kaaba? For some, the road is from Rum, for some from Syria, for some from Persia, for some from China, for some by sea from India and Yemen. So if you consider the roads, the variety is great and the divergence infinite; but when you consider the goal, they are all of one accord and one. The hearts of all are as one upon the Kaaba. The hearts have one attachment, an ardour and a great love for the Kaaba, and in that there is no room for contrariety. That attachment is neither infidelity nor faith; that is to say, that attachment is not confounded with the various roads which we have mentioned. Once they have arrived there, that disputation and war and diversity touching the roads—this man saying to that man, "You are false, you are an infidel" and the other replying in kind—once they have arrived at the Kaaba, it is realised that that warfare was concerning the roads only, and that their goal was one...[35]

References

1. The question: what is mysticism? is an issue in itself. For purposes of this paper it means, "the doctrine or belief that direct knowledge of God, of spiritual truth, of ultimate reality, etc., is attainable through immediate intuition, insight, or illumination, and in a way differing from ordinary sense perception or ratiocination" (*Webster's New International Dictionary* [second edition] Springfield, Mass., G & C. Merriam Co.) p.1621.
2. William James, *The Varieties of Religious Experience* (New York: The Modern Library, 1929) p.410. This passage is also quoted with approval in defence of the unitive nature of mysticism by T.M.P. Mahadevan, *Outlines of Hinduism* (Bombay: Chetana Ltd., 1971) p.4, who adds that this mystical "religious experience is the same all the world over. It is the crowning glory and consummation of man's existence on earth" (ibid). To move back now to the quotation from William James, it is elaborated at some length thus by James: "That are Thou! say the *Upanishads*, and the *Vedantists* add: 'Not a part, not a mode of That, but identically That, that absolute Spirit of the World'. 'As pure water poured into pure

water remains the same, thus, O Gautama, is the Self of a thinker who knows. Water in water, fire in fire, ether in ether, no one can distinguish them: likewise a man whose mind has entered into the Self'. Every man , says the Sufi Gulshan-Raz, 'whose heart is no longer shaken by any doubt, knows with certainty that there is no being save only One.... In his divine majesty the *me,* and *we,* the thou, are not found, for in the One there can be no distinction. Every being who is annulled and entirely separated from himself, hears, resound outside of him, this voice and this echo: *I am God:* he has an eternal way of existing, and is no longer subject to death'. In the vision of God, says Plotinus, 'what sees is not our reason, but something prior and superior to our reason... He who thus sees does not properly see, does not distinguish or imagine two things. He changes, he ceases to be himself, preserves nothing of himself. Absorbed in God, he makes but one with him, like a centre of a circle coinciding with another centre'. 'Here', writes Suso, 'the spirit dies, and yet is all alive in the marvels of the Godhead...and is lost in the stillness of the glorious dazzling obscurity and of the naked simple unity. It is in this modeless *where* that highest bliss is to be found', 'Ich bin so gross als Gott', sings Angelus Silesius again, 'Er ist als ich so klein. Er kann nicht uber mich, ich unter ihm nicht sein' (op. cit., pp.410-411). This extended extract is important in view of the fact that later on William James seems to qualify and tone down his statement considerably (ibid., pp.415-417).

3. S. Radhakrishnan, *Eastern Religions and Western Thought* (Oxford University Press, 1940) pp. viii-ix. Thus, "D.T. Suzuki in his *Mysticism, Christian and Buddhist* (1957) suggested an essential unity of experience as between selected Christian mystics, the Zen experience of *satori* ('enlightenment') and life lived in terms of Pure Land Buddhist invocation to Amitabha Buddha" (Michael Pye, *Comparative Religion* [Devon: David & Charles, 1972] p.110). Michael Pye later on also points out some limitations of this approach, as also of S. Radhakrishnan who writes elsewhere, "All the paths of ascent lead to the mountain top. This convergent tendency and the remarkable degree of agreement in the witness of those who reach the mountain top are the strongest proof of the truth of religion" (Robert A McDermott, ed., *Basic Writings of S. Radhakrishnan* [Bombay: Jaico Publishing House, 1970] p.290).
4. Frithjof Schuon, *Logic and Transcendence* (New York: Harper & Row 1975).
5. As for instance, Mircea Eliade would: *Patterns in Comparative Religion* (London: Sheed and Ward, 1958) passim. It is interesting to note, however, that in the Indian context, Mirceal Eliade does seem to see Yoga-mysticism as a unifying factor in Hinduism quite markedly (see Mircea Eliade, *Yoga* [New York: Pantheon Books, 1954] Chapter Nine). This approach may be contrasted with that of Surendranath Dasgupta for whom Yoga-mysticism seems to be one form of several types of Hindu mysticism. See S.N. Dasgupta, *Hindu Mysticism* (New York: Frederick Ungar Publishing Co., 1959) passim. For various types of Hindu mysticism in the RigVedic period, see T.F. Mainkar, *Mysticism in the RgVeda* (Bombay: Popular Books Dept., 1961).
6. See M.K. Gandhi, *The Story of My Experiments with Truth* (Washington, D.C.: Public Affairs Press, 1940) p.207, etc.
7. Some scholars, however, do not distinguish between these levels in emphasizing the unity of religions (see Bhagavan Das, *The Essential Unity of All Religions* [Benares: The Indian Bookshop, 1947] passim). Others who emphasize the mystical element go so far as to posit it as an *ens*. Thus Ben-Ami Scharfstein says

of mysticism that its "first quintessence is sameness. Even the mystic confesses that things look different from one another, but their difference, he says, is illusory. There are not really, he says, any stones, butterflies, raindrops, cats, trees, dollars, houses, Hindus, Jews, persons called Luigi, Hayyim, margot, or even you or I; or, rather, they are all the same one thing, which is different from all their differences" (*Mystical Experience* [Oxford: Basil Blackwell, 1973] p.142). He also discusses the several techniques elaborated in mysticism to establish this (ibid., pp.142-146).

8. Gerschom G. Scholem, *Major Trends in Jewish Mysticism* (New York: Schocken Books, 1974) pp.5-6. Part of it is also quoted in Robert Lawson Slater, '*World Religions and World Community* (Columbia University Press, 1963) p.8. It may be noted, however, that Scholem also adds, "That there remains a common characteristic, it would be absurd to deny, and it is this element which is brought out in the comparative analysis of particular mystical experiences" (op.cit.,p.6).

9. Consider for instance, this intensely Christian mystical experience of Pascal whose account was found sewn to the inside of his coat after his death:

"In the year of Grace, 1654,
On Monday, 23rd of November, Feast of St. Clement,
Pope and Martyr, and of others in the Martyrology,
Vigil of Saint Chrysogonus, martyr, and others,
From about half-past ten in the evening until about half-past twelve."

FIRE

God of Abraham, God of Isaac, God of Jacob, not of the
philosophers and scholars.
Certitude Certitude. Feeling. Joy. Peace
God of Jesus Christ.
My God and thy God.
"Thy God shall be my God".
Forgetfulness of the world and of everything, except God.
He is to be found only by the ways taught in the Gospel.
Greatness of the soul of man.
"Righteous Father, the world hath not known Thee, but I have known Thee".
Joy, joy, joy, tears of joy.
I have fallen away from Him.
They have forsaken Me, the Fountain of living waters.
"My God, wilt thou forsake me?"
May I not fall from Him for ever.
"This is life eternal, that they might know Thee, The only true
God. and Jesus Christ, Whom Thou hast sent".
Jesus Christ.
Jesus Christ.
I have fallen away: I have fled from Him, denied Him,
Crucified Him.
May I not fall from Him for ever.
We keep hold of Him only by the ways taught in the Gospel.
Renunciation, total and sweet.
Total submission to Jesus Christ and to my director.
Eternally in joy for a day's exercise on earth.

"I will not forget Thy word. Amen".
See Emily Cailliet, *The Clue to Pascal* (London: S.C.M. Press, 1944) pp.47-48.

10. *The New Encyclopaedia Britannica,* Macropaedia Vol.12 (Chicago: William Benton, 1974) p.787.
11. R.C. Zaehner, *Mysticism Sacred and Profane* (Oxford: Clarendon Press, 1957) passim; also see *Drugs, Mysticism, and Make-believe* (London: Collins, 1972). Michael Pye (op.cit., p.109) cities R.C. Zaehner's *Mysticism Sacred and Profane* as an "analysis of mysticism... controlled by theological criteria of doubtful relevance".
12. R.C. Zaehner, *Hindu and Muslim Mysticism* (London: Athlone Press, 1960).
13. R.C. Zaehner, *The Bhagavad Gita* (Oxford: Clarendon Press, 1965) pp.2-3.
14. Ibid., p.2. It may be pointed out, however, that R.C. Zaehner accepts the fact, though rare, of some mystics having both the experiences and witnessing the distinction between the two. In this connection he cites the 'Flemish Ryusbroeck' and then "the Islamic mystics, Al-Junayd of Baghdad, Ibn Tufayl of Andalusia, and the Iranian Najm al-Din Razi", and on the Hindus side, "pre-eminently" the Bhagavad Gita (ibid., pp.2-3).
15. See Mariasusai Dhavamony, *Love of God According to Saiva Siddhanta* (Oxford: Clarendon Press, 1971) pp. 367-374. It is interesting to note that Rudolf Otto felt justified in "speaking of uniform nature of mysticism" on account of "an astonishing conformity in the deepest impulses of human spiritual experience"; yet he drew distinctions among types of mysticism *both* on the basis of historical differences in tradition and *internal* differentiation of mysticism into various types. After differentiating mystical experience by means of Indian examples he concludes, "Christian analogies might be found for all or almost all the above-mentioned forms of Indian Mysticism. Together with the latter they would prove that mysticism admits of wide variations and they would also serve to show how in spite of "convergence of types' between East and West, the inner spirit yet differs, and that the very different ground upon which mysticism rose in Europe also colours the highest mystical experience in a way that is Christian and not Indian" (Rudolf Otto, <u>Mysticism East and West</u> [Toronto: The Macmillan Co., 1960] p.179).
16. R.C. Zaehner, *Mysticism Sacred and Profane,* p.204.
17. Ibid., pp.198-199. Also see "Interpretation and Mystical Experience" by Ninian Smart, *Religious Studies*, Vol.1-2 (1966-67) p.76. This Nature Mysticism (or what R.C. Zaehner likes to call pan-en-henism) should be distinguished from the use of the term "natural mysticism" by Heinrich Dumoulin by which he means "an immediate religious experience of reality or a psychic contact with absolute being" (*A History of Zen Buddhism* [Boston: is Beacon Press, 1963] p.282). This is more like what R.C. Zaehner refers to as monistic type of mysticism. Similarly Dumoulin uses the term "supernatural mysticism" and the expression "supernatural mysticism of grace" (ibid., p.283). He writes: "Christian mysticism is thus a matter of *grace* and is essentially supernatural. As a supernatural manifestation it belongs to higher order than natural mysticism" (ibid., p.283). Michael Pye is quick to note the value judgement involved here (op.cit., p.110) but what interests us is the parallel his distinctions provide to the monistic-theistic type of mysticisms identified by R.C. Zaehner. For all practical purposes then, Dumoulin may be subsumed under Zaehner's distinction, except for the expressions used to designate them, which are a trifle misleading.

18. See, for example, R.C. Zaehner, *Mysticism Sacred and Profane*, p.198.
19. It is interesting to speculate on some of the factors that might explain this turn-around in the understanding of the nature of mysticism. This has largely been a Western academic development, for the Hindu tradition, both academic and religious, continues to stand by, it seems, the earlier view which emphasises mystical unity. Some of the factors operating in the West which might account for this change could be a change in the academic climate in which the study of religion is carried out. Robert Lawson Slater remarks: "Whereas pioneers in the field, a century ago, were convinced that further study would demonstrate the fundamental agr eement of apparently dissimilar statements, a good many today are disposed to draw precisely the opposite conclusion from this same further study: apparently similar statements, they affirm, when interpreted in the light of their own particular contexts, are seen to be essentially dissimilar" (op.cit., p.8; also see p.231 note 11). Martin Buber's influence may also in part account for this change when he is taken as representative of those to whom the *Advaitic* "undifferentiated unity or union between the individual and the supreme self is unacceptable..." The Jewish philosopher Martin Buber emphasized an "I-Thou" relationship: 'All real living is meeting', and one Thou cannot become It. But even his own 'unforgettable experience' of union he would explain as 'illusory' *(The New Encyclopaedia Britannica Macropaedia*, Vol. 12, p.787). The shift away from mystical universalism may also partly be due to the fact that "monism of the Hindu type seems now less popular with Western Philosophers than it was in the last century" (R.C. Zaehner, ed., *The Concise Encyclopaedia of Living Faiths* [Boston: Beacon Press, 1972] p.260).
20. See Christopher Isherwood, *Ramakrishna and His Disciples* (London: Methuen & Co. Ltd, 1965) pp.1-2. Aldous Huxley 'Preface' to Swami Nikhilananda, *The Gospel of Sri Ramakrishna* (New York: Ramakrishna-Vivekananda Centre, 1942); Swami Saradananda, *Sri Ramakrishna The Great Master* (Madras: Sri Ramakrishna Math, 1952).
21. Swami Sardananda, op. cit., p.295.
22. Ibid., p.296. The account continues, "Jesus, the god-man, then embraced the Master and disappeared into his body and the master entered into ecstasy, lost normal consciousness and remained identified for some time with the Omnipresent *Brahman* with attributes. Having thus attained the vision of Jesus, the Master became free from the slightest doubt about his being an incarnation of God" (ibid. Emphasis supplied).
23. Ibid., p.260. The account continues, "afterwards he had the knowledge of the all-pervading *Brahman* with attributes and merged finally in the attributeless *Brahman*, the Absolute" (ibid. Emphasis supplied).
24. Swami Ghanananda, *Sri Ramakrishna and His Unique Message* (London: The Ramakrishna Vedanta Centre, 1970) p.27.
25. ibid., p.61. Emphasis supplied.
26. ibid., pp.85-87. Emphasis supplied.
27. ibid., p.478. Emphasis supplied.
28. Swami Nikhilananda, op.cit., pp.237, 478, 639, 702, 812, etc.
29. It is surprising that R.C. Zaehner does not mention Ramakrishna not only among those who have made the distinction but also among those "very few" mystics "who had had both experiences and had witnessed the distinction between the two". The reason is not far to seek. According to R.C. Zaehner, Ramakrishna "was by nature

a *bhakta* [sic], not a *jnanin*" (*Hinduism*, Oxford University Press, 1962, p.165). His failure to perceive Ramakrishna as bearing witness to both *bhakti* and *jnana* seems to have seriously handicapped his analysis of the distinction between the two types of mysticism. By this misapprehension, he ruled out from consideration the richness of biographical and autobiographical detail we possess about Ramakrishna, to the "fidelity and wealth" of which Aldous Huxley draws attention, as well as to the uniqueness of the fact that such firsthand accounts should be available about a mystic and that of the stature of Ramakrishna (Swami Nikhilananda, op.cit., p.ii). For evidence of the fact that Ramakrishna had both types of mystical experiences referred to above, see Swami Saradananda, op.cit., chapters VI-XV, especially pp.140-141 and 251-252. Ramakrishna is also very explicit in regarding both types of mysticism as representing the experience of the same Reality. Ramakrishna told Goswami, "How can you say that the only truth about God is that He has form. It is undoubtedly true that God comes down to earth in human form, as in the case of Krishna. And it is true as well that God reveals himself to His devotees in various forms. But it is also true that God is formless. He is the Indivisible Existence-Knowledge-Bliss Absolute. He has been described in the Vedas both as formless and as endowed with form. He is also described there both as attributeless and as endowed with attributes" (Swami Nikhilananda, op. cit., p.191). A member of the Brahmo Samaj which accepted "God without form but with attributes" (Nalini Devdas, *Sri Ramakrishna*, [Bangalore: The Christian Institute for the Study of Religion and Society, 1966] p.32), Pratap Chandra Majumdar described him thus in the Theistic Quarterly Review: "...He worships Siva, he worships Kali, he worships Rama, he worships Krishna, and is a confirmed advocate of Vedantic doctrines... He is an idolator, yet is faithful and most devoted meditator on the perfections of the One Formless, Absolute, Infinite Deity" (ibid., p.45).

30. Swami Nikhilananda, op. cit., p.191. Emphasis supplied.
31. Romain Rolland, *The Life of Ramakrishna* (Calcutta: Advaita Ashrama, 1960) p.184, fn.2.
32. To get a full idea of his versatility in this respect, see Claude Alan Stark, *God of All* (Cape Cod, Mass.: Claude Stark, Inc., 1974).
33. See Huston Smith, *The Religions of Man* (New York: Harper & Row, 1958) pp.86-87.
34. See Romain Rolland, op. cit., p.273, n.2.
35. Quoted in Michael Pye, op.cit., pp.150-151. For the possibility that Ramakrishna's attitude to other religions and practices may be even *more* tolerant than is usually depicted, see Walter G. Neevel, Jr., "The Transformation of Sri Ramakrishna" in Bardwell L. Smith, *Hinduism: New Essays in the History of Religions* (Leiden: E.J. Brill, 1976) p.76.

RAMAKRISHNA ON THE NATURE OF *BHAKTI* AND *JNANA*

I

From the time of his earliest biographers, the question whether Ramakrishna was a *bhakta* or a *jnanin* has been a troublesome one to answer. Thus one of his earliest biographers, Max Mueller, wrote in the Preface to his work that "the fervent love of God, nay the Godhead, has nowhere found a stronger and more eloquent expression than in the utterances of Ramakrishna, yet he perfectly knew the barriers that separate divine and human nature".[1] This is somewhat equivocal but elsewhere in the book Max Mueller expresses himself more definitely when he points out that Ramakrishna,

> ... himself distinguishes very clearly between philosophy or *jnana* (knowledge) and devotion or *bhakti*, and he himself was a *Bhakta*. or a worshipper or lover of the deity, much more than a *jnanin* or a knower.[2]

That Ramakrishna was pre-eminently a *bhakta* is accepted by most biographers.[3] The problem arises when one wishes to consider whether this fact refers to his temperament or to a relative assessment of *bhakti* and *jnana* as avenues to Reality. In the first case his being a *bhakta* becomes a matter of personal preference, in the other case it is a matter of general concern: did he merely *prefer bhakti* or did he regard it as *superior* to *jnana?*

II

There are several utterances of Ramakrishna which can be cited in support of the view that he regarded *bhakti* as "greater" than *jnana*. Thus he said:

> One can attain everything through *bhaktiyoga*, I wept before the Mother and prayed, 'O Mother, please tell me, please reveal to me, what the *yogis* have realized through *yoga* and the *jnanis* through discrimination'. And the Mother has revealed everything to me. She reveals everything if the devotee cries to her with a yearning heart. She has shown me everything that is in the *Vedas*, the *Vedanta*, the *Puranas* and the *Tantra*.[4]

He also told his devotees:

> *Bhakti* may be likened to a woman who has access to the inner court of a house. *Jnana* can only go as far as the outer rooms.[5]

This statement was also sent by Swami Vivekananda to Max Mueller as a part of the Sayings of Ramakrishna. It occurs there as saying # 172 thus.

> Thy Knowledge of God may be likened to a man, while the Love of God is like a woman. Knowledge has entry only up to the outer rooms of God, but no one can enter into the inner mysteries of God save a lover, for a woman has access even into the harem of the Almighty.[6]

III

On the other hand, however, there are statements made by Ramakrishna and metaphors used, which seem to place *jnana* in a superior position.[7] He compares the achievement of *jnana* to the highest note in the scale of music,[8] to the roof,[9] to a mountain.[10] If the point is phrased in terms of Shankara's *Advaita* and Ramanuja's *Visistadvaita* then the following account is of interest. Two of his disciples were disputing over the respective positions of the two schools, one of them the would-be Vivekananda, when Sri Ramakrishna said to M (the author of the *Gospel of Sri Ramakrishna*):

> I do not enjoy these discussions. Why should I argue at all? I clearly see God in everything, He himself has become all. I see that

whatever is, is God. He is everything; again, He is beyond everything. I come to a state in which my mind and intellect merge in the indivisible... I cannot utter a word unless I come at least two steps from the plane of *Samadhi*. Sankara's non-dualistic explanation is true, and so is the qualified non-dualistic interpretation of Ramanuja.[11]

IV

How then are these apparent contradictions to be reconciled, short of reiterating Ramakrishna's statement that "Knowledge and love of God are ultimately one and the same. There is no difference between pure knowledge and pure love."[12]

Though it seems that through one the other may be obtained, it is clear that according to Ramakrishna, the hallmark of *bhakti* is *intimacy* and of *jnana*, *ultimacy*.

References

1. Max Mueller, *Ramakrishna: His Life and Sayings* (London: Longmans, Green & Co., 1898) pp.ix-x.
2. Ibid., pp.93-94. Spellings modified.
3. "I have told you that the devotee who is born with an element of Vishnu cannot altogether get rid of *bhakti*. Once I fell into the clutches of a *jnani*, who made me listen to *Vedanta* for eleven months. But he couldn't altogether destroy the seed of *bhakti* in me. No matter where my mind wandered it would come back to the Divine Mother" (Swami Nikhilananda (tr.) *The Gospel of Sri Ramakrishna* [Madras: Sri Ramakrishna Math, 1964] p.759). This obviously refers to his meeting Tota Puri. When Ramakrishna was initiated into Advaita Vedanta by Tota Puri, the vision of Kali became a hindrance to the achievement of *nirvikalpa samadhi*, whereupon Ramakrishna says that he "looked upon knowledge as a sword and cut it mentally in two with the sword of knowledge. There remained then no function in the mind, which transcended quickly the realm of names and forms, making me merge in *Samadhi* (Swami Saradananda, *Sri Ramakrishna: The Great Master* [Madras: Sri Ramakrishna Math, 1952] p.251). He also refers to remaining in the *nirvikalpa samadhi* of the *jnani* for six months, "afterwards the Mother's command was heard, 'Remain in *Bhavamukha*, for the spiritual

enlightenment of the people remain in *Bhavamukha*" (ibid.) *Bhakti* and *jnana* then reflect different phases of life as well.

4. Swami Nikhilananda (tr.) *The Gospel of Sri Ramakrishna* p.542. Emphasis supplied.
5. Ibid., p.844. Interestingly, Dr. Sarkar, who was attending on him commented, "All women are not allowed to enter the inner court; for instance, prostitutes. Hence the need for *jnana*" (ibid.).
6. Max Mueller, op.cit., p.138.
7. See Swami Ramakrishnanda, *Life of Sri Ramanuja* (Madras: Sri Ramakrishna Math, 1959) p.i.
8. Swami Nikhilananda (tr.), op.cit.,30,437.
9. Ibid., p.30.
10. Ibid.
11. Swami Ramakrishnananda, op.cit., p.i. Emphasis supplied.
12. Max Mueller, op.cit., p.ix.

4

RAMAKRISHNA PARAMAHAMSA — *JNANIN* OR *BHAKTA?* PATTERNS IN RECONCILIATION

I

There are several scholars who are inclined to regard Ramakrishna primarily as a *jnanin*.[1] There are also several other scholars who are inclined to regard him primarily as a *bhakta*.[2] No sensitive scholar can ignore the fact, however, that if Ramakrishna is regarded only as a *jnanin*, a considerable body of evidence which attests to the contrary has to be satisfactorily explained.[3] Similarly no conscientious scholar who studies the life of Ramakrishna regarding him as a *bhakta* can ignore the fact that a considerable body of evidence points in the direction of his being a *jnanin*.[4]

This essay attempts to examine the patterns of reconciliation employed by hagiographers, biographers and scholars to accommodate the divergent pieces of evidence on this issue.

II

Two disciples of Ramakrishna seem to regard Ramakrishna primarily as a *jnanin*: Swami Vivekananda and Swami Saradananda. They seem to explain the apparent emphasis on *bhakti* in the teachings of Ramakrishna in terms of his *upayakausalya*. Thus Swami Vivekananda explicitly states:

Devotion as taught by Narada, he used to preach to the masses, those

> who were incapable of any higher training. He used generally to teach dualism. As a rule, he never taught *Advaitism.* But he taught it to me. I had been a dualist before.[5]

Ramakrishna's teaching on this aspect, then, varied according to the person involved as he "used to see that different people required different treatment."[6] Thus the answer to the question whether Ramakrishna was a *jnanin* or *a bhakta* depends not only on whom we are talking to; it might also depend on whom Ramakrishna was talking to! Vivekananda clearly speaks also of an "inner circle"[7] of Ramakrishna with which he *may* have shared insights different from those he shared with others.

Similarly Swami Saradananda remarks:

> ...although he was of the opinion that *Advaita* knowledge was the ultimate truth, he always taught the general public with its hankering for worldly objects, the truth of qualified monism, and not unoften, the love of God after the manner of a dualist.[8]

Thus one pattern of reconciling the two strands in Ramakrishna's teaching—of *jnana* and *bhakti*—can now be identified. *Jnana* constituted the esoteric message, *bhakti* the exoteric one.

III

R.C. Zaehner is one of those scholars who regard Ramakrishna as a *bhakta* and not a *jnanin.* While those who regarded Ramakrishna as a *jnanin* explained Ramakrishna's talk of *bhakti* in terms of the temperament of his disciples; R.C. Zaehner, who regards him as a *bhakta,* explains this position in terms of Ramakrishna's *own* temperament. Thus he writes:

> Ramakrishna was by nature a *bhakta* [sic], not a *jnanin*, a worshipper of God rather than an introspective seeking the One within himself. Despite his *Advaitin* teacher, however, his trances did not reveal a static Monad, One without a second, they showed rather that 'God has become all things.' Intellectually he may deny this on occasion but Ramakrishna set no store by the intellect.[9]

Thus this second pattern of reconciliation of the two strands in

Ramakrishna's teaching—of *jnana* and *bhakti*—reconciles them by regarding the former as representing an external influence and the latter a personal preference.

IV

Another pattern of reconciliation adopted by some scholars is not pedagogical as in the first case, nor personal as in the second, but rather historical. It is pointed out that while Ramakrishna was really a *bhakta*, he was made out into a *jnanin* by his disciples.[10] Thus R.C. Zaehner says that though Ramakrishna was a *bhakta*, "his followers, however, developed philosophically in a purely monistic direction".[11] K.M. Sen, too, thinks that while Ramakrishna "preached pure *bhakti* without theological complications", his disciple, Swami Vivekananda "spread this Vedantic doctrine of *advaita* (non-dualism) in India and abroad."[12] Unlike some views which recognise both *bhakti* and *jnana—oriented* elements in the experiences of Ramakrishna himself, on this view the *jnana* element is seen as more or less a posthumous graft on Ramakrishna. There is, however, enough biographical evidence to show that Ramakrishna himself had the experience of *jnana*, and hence this view is hard to support, though it may be consistent with the fact that the followers of Ramakrishna chose to emphasise the *jnana* element of his experience.

V

One further pattern can be traced through which attempts are made to reconcile these two aspects—*jnana* and *bhakti*—of Ramakrishna's life and teachings. On this view, the key lies not in the nature of his following, or in his own nature or that of his followers but in his voracious appetite for spiritual experience. The explanation of the co-existence of the elements of *jnana* and *bhakti* in him is neither pedagogical, nor temperamental or apostolic but simply spiritual. This spiritual versatility was such that "In a twelve year period he meditated like a *yogi*, worshipped like a *bhakta*, practised Jainism, Buddhism and Shaktism, and experienced the reality both of *Brahman* without attributes (the impersonal *Brahman*) and *Brahman* with attributes (i.e. the personal God). Dressed like a Muslim, he prayed until he knew God as Allah, he turned to Christianity and found God in Christ."[13]

To conclude: writers on Ramakrishna have tried to reconcile the presence of the elements of both *jnana* and *bhakti* in his experience and teachings in terms of his pedagogy, his personality, his apostles, and his multi-faceted spirituality. On the whole, while all of the points seem to possess some force, the most cogent explanation of the simultaneous presence of elements of *jnana* and *bhakti* seems to lie in his spiritual versatility.

References

1. See Harold W. French, *The Swan's Wide Waters* (London: Kennikat Press, 1974) pp. 30-31; T.M.P. Mahadevan, *Outlines of Hinduism* (Bombay: Chetana Ltd., 1956) p.222; Haridas Bhattacharyya ed., *The Cultural Heritage of India* (Calcutta: The Ramakrishna Institute of Culture, 1956) p. 674; V.S. Narvane, *Modern Indian Thought* (New York: Asia Publishing House, 1964) Chapter III; P.T. Raju, *The Philosophical Traditions of India* (University of Pittsburg Press, 1971) p. 230; Swami Saradananda, *Sri Ramakrishna, The Great Master* (Madras: Sri Ramakrishna Math, 1952) p. 704; R.C. Zaehner, ed., *The Concise Encyclopaedia of Living Faiths* (Boston: Beacon Press, 1959) p. 257 and perhaps also Wm. Theodore de Bary, *Sources of Indian Tradition* Vol.II (New York: Columbia University Press, 1958) p.85. The portrayal of Ramakrishna as an *Advaitin* may also in part be due to the influence of his famous disciple Swami Vivekananda, Amaury de Riencourt, *The Soul of India* (London: Jonathan Cape, 1961) p.243; K.M. Sen, *Hinduism* (Baltimore: Penguin Books, 1973 [first published 1961]) p.110; etc.
2. See F. Max Mueller, *Ramakrishna: His Life and Sayings* (Bombay: Longmans, Green & Co., 1898) p. 94; R.C. Zaehner, *Hinduism* (Oxford University Press, 1966) p. 165; Thomas J. Hopkins, *The Hindu Religious Tradition* (Belmont, California: Dickenson Publishing Co., 1971) p.137; S.G.F. Brandon, *A Dictionary of Comparative Religion* (New York: Charles Scribners Sons, 1970) p. 529; Arthur Osborne, *Buddhism and Christianity in the Light of Hinduism* (London: Rider & Co., 1959) p. 52; Agehananda Bharati, *The Ochre Robe* (London: George Allen & Unwin Ltd., 1961) p. 220; etc.
3. See Swami Saradananda, op. cit., passim; Swami Nikhilananda, *The Gospel of Sri Ramakrishna* (Madras: Sri Ramakrishna Math, 1964); Romain Rolland, *The Life of Ramakrishna* (Calcutta: Advaita Ashrama, 1960); Christopher Isherwood, *Ramakrishna and his Disciples* (London: Methuen & Co., 1965); Claude Alan Stark, *God of All* (Cape Cod, mass.: Claude Starke, Inc., 1974); etc.
4. See previous note.
5. *The Complete Works of Swami Vivekananda* Vol. VII (Calcutta: Advaita Ashrama, 1969) p. 412. Emphasis supplied.

6. Ibid., p.412 Cf. "As a mother, in nursing her sick children, gives rice and curry to one, and sago arrowroot to another, and bread and butter to a third, so the Lord has laid out different paths for different men suitable to their natures" (quoted in Huston Smith, *The Religions of Man* [New York: Harper & Row, 1958] p.87). 7. *The Complete Works of Swami Vivekananda,* Vol. VII, pp. 411-412.
8. Swami Saradananda, op. cit., p. 704. Emphasis supplied.
9. R.C. Zaehner, op. cit., p. 165. Third emphasis supplied. The last statement is true but can cut both ways. Swami Saradananda tells us that "He used to be greatly annoyed at persons who, having no high spiritual experience, not even love of God, would, nevertheless bandy arguments in favour of non-dualism or qualified non-dualism" (op.cit., 704).
10. One must contend here with Vivekananda's apple-cart-upsetting remarks about Ramakrishna: "Outwardly he was all *bhakta,* but inwardly all *jnanin* ... I am the exact opposite" (quoted by Amaury de Riencourt, *The Soul of India* (London: Jonathan Cape, 1961] p. 243).
11. R.C. Zaehner, op.cit., p. 165.
12. K.M. Sen, op. cit., p. 110.
13. John B. Noss, *Man's Religions* (New York: The Macmillan Co., 1967) pp. 300-301. Emphasis added. The case for Ramakrishna having practised Jainism does not seem to be as strong as Noss suggests (see Swami Saradananda, op. cit., p.297). For detailed account of Ramakrishna's spiritual versatility see Claude Alan Stark, op. cit., passim.

5

SRI RAMAKRISHNA AND SWAMI SHANKARACARYA : A COMPARISON

I

Ramakrishna (1836-1886 AD)[1] and Shankara (788-820 AD)[2] are two great names in the history of Hinduism, and the temptation to institute at least a passing comparison between them must have been felt by many a student of Hinduism.[3] The reason why such a comparison has not been carried out in any detail is not far to seek. It seems to lie in the poverty of genuine information about Shankara's life (which contrasts sharply with the much more detailed information we possess about his philosophy) and the wealth of detail we possess about Ramakrishna's life and experiences. There is little information about Shankara which is not legendary[4] and sometimes the legends are at variance;[5] moreover, the question as to which works Shankara actually wrote, as distinguished from those which are ascribed to him, still remains to be definitely resolved.[6] By contrast, although, unlike Shankara, Ramakrishna did not write philosophical works, "Ramakrishna's life, being comparatively recent history, is well documented. In this respect, it has advantage over the lives of other, earlier phenomena of a like nature. We do not have to rely, here, on fragmentary or glossed manuscripts, dubious witness, pious legends."[7]

There are thus basic differences in the nature of the evidentiary material, on the basis of which alone a genuine comparison can be instituted. One is the difference in the amount of evidence available to us, which is meagre in the case of Shankara and substantial in the case of Ramakrishna. Another difference lies in the nature of its quality—in one case it is legendary, in the other based on eyewitness accounts, "based on words and deeds Ramakrishna undubitably spoke and did."[8] Yet another difference pertains to the nature of that evidence. In the case of Shankara most of this information is philosophical; in the case of Ramakrishna it is biographical. Thus if one did institute a comparison on the basis of trustworthy evidence about both, one would end up comparing a *philosophy with a person.* This involves a slight overstatement as philosophical utterances of Ramakrishna are fully attested to but there can be little doubt that the main factor in Shankara's influence has been his philosophy (though he seems to have led a vigorous life),[9] and the main factor in Ramakrishna's influence has been his life (though he did on occasion make statements with clear philosophical implications). Notwithstanding these limitations, however, an effort will now be made to compare the relationship between the philosophies and experiences of Shankara and Ramakrishna.

II

It is obvious that before such a comparison can be instituted, a reconstruction of the spiritual life of Shankara is necessary on the basis of the traditional account. Once this has been accomplished, the desired comparison could then be carried out. In this task help must be sought, not from his dialectical *Digvijayas* but from his own works. Two key bits of information about his personal spiritual pilgrimage can be garnered from these works.

The first derives from his gloss of *Vedantasutra* IV.1.15. The passage runs as follows: *katham hi ekasya sva-hrdayapratyayam brahmavedanam deha-dharanam ca-aparena pratikseptum sakyeta?* T.M.P. Mahadevan, who quotes this passage, remarks that Shankara:

> ...set the model for subsequent metaphysical thinking, and gave a new direction to the course of philosophical history in India. This he was able to do because of a conviction born of the highest experience, as is evidenced by the only reference—even that, indirect—he makes to his own realization towards the end of his *sutra-bhasya*. The subject he discusses here is *jivan-mukti* (liberation while being embodied). In this context he asks, 'How can one contest the heart-felt cognition of another as possessing *Brahman*-knowledge, even though bearing a body?' Shankara has poured

forth, for the benefit of posterity, great truths from the depths of his realization ...[10]

M. Hiriyanna, again while talking of the *jivanmukta*, points out that a "*jivan-mukta* experiences pain and pleasure, but neither really matters to him. He does not necessarily give up all activity as is abundantly illustrated by the strenuous life which Shankara himself led..."[11] At this point, he adds a significant footnote in which he mentions that "It is interesting in this connection to refer to Shankara's statement at the end of his com. on VS.IV.i.15, which tradition views as an allusion to his own direct experience of the ultimate truth".[12]

This may then provisionally be accepted as a reference to Shankara's own realization of *(nirguna)* Brahman.

The second key bit of information may be garnered from the numerous devotional hymns that are ascribed to Shankara.[13] These include such hymns as the *Daksinamurti Stotra, Harimide Stotra, Anandalahari* and *Saundaryalahari*.[14] Western scholars in general tend to reject Shankara's authorship of these hymns [15] but the tradition is so strong [16] that one may at least accept the fact of Shankara having composed devotional hymns, even if his authorship of the existing hymns cannot be established.

III

Now one is, prima facie, better prepared to compare Shankara and Ramakrishna. It may be assumed that Shankara's religious experience had the dimensions of *jnana*, represented by his claim of realization of *Brahman*, and *bhakti* as represented by his devotional hymns.[17] The same holds true of Ramakrishna.[18] There is, however, one difficulty. We know the sequence in which Ramakrishna has his experiences. He first had devotional experiences, then *advaitic* experiences and this was again followed by *bhaktic* experiences. In the case of Shankara one tends to presume that the *advaitic* experience preceded the *bhaktic* but there seems to be no way of establishing this. Nevertheless if it be accepted that Shankara had both *advaitic and bhaktic* experiences, something we know for a fact about Ramakrishna, then we are in a position now to ask a comparative question. What is the relation of the philosophical positions of Shankara and Ramakrishna to their personal spiritual experiences?

IV

In order to answer this question it is important to realize that the philosophical positions of Shankara and Ramakrishna differ. Shankara is an exponent of *Advaita,*[19] but although Ramakrishna had *Advaita* experiences he was not an exponent of *Advaita* as such. This calls for some elaboration. Let it first be established that Ramakrishna did have the *advaitic* experience.

The first major devotional experience of Ramakrishna was a vision of *Kali.*[20] Thereafter, when he was initiated into *Advaita Vedanta* by Totapuri, the same vision of Kali became a hindrance to the achievement of *nirvikalpa samadhi,* the ultimate experience in *Advaita.* In Ramakrishna's own words, "The mind withdrew itself from all other things but, as soon as it did so, the intimately familiar form of the universal Mother, consisting of the effulgence of pure consciousness, appeared before it as living and moving and made it quite oblivious of the renunciation of names and forms of all descriptions".[21] In order to finally achieve *nirvikalpa samadhi* in which *nirguna brahma* is experienced, when "the holy form of the divine mother appeared now before the mind as previously," Ramakrishna "looked upon knowledge as a sword and cut it mentally in two with the sword of knowledge. There remained then no function in the mind, which transcended quickly the realm of names and forms, making me merge in *Samadhi.*"[22] This trance lasted three days, later Ramakrishna was in that state continually for six months. "Then afterwards, the Mother's command was heard, "Remain in Bhavamukha'"[23] or in *savikalpa samadhi*, a step below *nirvikalpa samadhi*[24]. Thus clearly Ramakrishna did have the "peak experience" of *Advaita.*

But now the next point. Although one may go up to the "top of the mountain", Ramakrishna does not advise living there forever. In his Philosophy Shankara allows for *jivanmukti* but would not, it seems, opt for the slopes of the mountain as Ramakrishna seems to:

> A man cannot live on the roof a long time. He comes down again. Those who realize *Brahman* in *samadhi* come down also and find that it is *Brahman* that has become the universe and its living beings. The ego does not vanish altogether. The man coming down from *samadhi* perceives that it is *Brahman* that has become the ego the universe, all living beings. This is known as *vijnana.*[25]

This is hard to accept in *Advaita Vedanta* for when one comes down, one has come down into the world of *saguna brahma* and even *maya* and "*saguna Brahman* includes not merely reality but also appearance, which is something less than the real. The element of reality in it is the ultimate of *Advaita*,"[26] from which "appearance must be excluded".

It is important to realize however, that although Ramakrishna's philosophy does not fully conform to Shankara's *Advaita*, if one probes the lives of these two figures one may find that there are closer biographical parallels between them than may appear at first sight. It has already been shown how Shankara was not only an *Advaitin* but may have composed many devotional hymns as well,[27] so much so that the legend (which seems to know everything when historians know nothing) informs us that "when his mother was dying, he recited to her first, a hymn of Siva and then one of Vishnu".[28] Obviously Shankara too came down from his "peak" experience of non-dual reality and often resorted to an attitude of devotion like Ramakrishna. The difference, however, cannot be altogether overlooked that while Shankara's teaching was perhaps developed entirely in the light of his peak experience, Ramakrishna preached not from the peak but from the slopes.

Shankara and Ramakrishna both came back to the world. But Shankara's philosophical position was "frozen" at the high altitude of the spiritual summit of *nirguna brahman*. He descended from the summit but not his philosophy. Ramakrishna came down too, and came down with an adequate philosophy for "coming down".

References

1. See T.M.P. Mahadevan, *Outlines of Hinduism* (Bombay: Chetana Ltd., 1971) p. 216; Thomas J. Hopkins, *The Hindu Religious Tradition* (Belmont, California: Dickenson Publishing Company, Inc., 1971) pp. 136-137; but also see R.C. Zaehner, ed., *The Concise Encyclopaedia of Living Faiths* (Boston: Beacon Press, 1959) p. 257; J.N. Farquhar, *Modern Religious Movements in India* (New York: MacMillan Co., 1915) p.188 for somewhat different dates.
2. See T.M.P. Mahadevan, op.cit., p. 141; but also see S. Radhakrishnan, *Indian Philosophy* (London: George Allen & Unwin Ltd., 1951) pp.447, 448 fn 2.
3. See R.C. Zaehner, *Hinduism* (Oxford University Press, 1966) p. 188. It seems, though, that Vivekananda perhaps lends himself more readily for being associated with Shankara than Ramakrishna (ibid., p.168 etc.).

4. See S. Ramakrishna, op.cit., pp.447-451; Benjamin Walker, *The Hindu World* (New York: Frederick A. Praeger, 1968) pp. 348-352; N.V. Raman, *Sankaracharya the Great* (Madras, 1923) passim; C.N.K. Aiyar, *Life and Times of Sankara* (Madras, 1934?) passim; Eliot Deutsch and J.A.B. van Buitenen, *A Source Book of Advaita Vedanta* (Honolulu: University of Hawaii Press, 1971) pp. 122-124.
5. Benjamin Walker, op. cit., p. 349.
6. Paul Hacker and Sengaku Mayeda are two scholars who have specially addressed themselves to this question.
7. Christopher Isherwood, *Ramakrishna and his Disciples* (London: Methuen & Co. Ltd., 1965) p.1.
8. ibid., p.1. The difference is highlighted when we compare the two main biographers of Shankara and Ramakrishna. In the case of Shankara the chief biographical accounts are Madhava's *Samkaradigvijaya* and Anandagiri's *Samkaravijaya* (see S. Ramakrishna, op.cit., pp.447-448). These are rather hagiographical. By contrast, the two chief authorities on Ramakrishna and Mahendra Nath Gupta's *The Gospel of Sri Ramakrishna* and Swami Saradananda's *Sri Ramakrishna the Great Master*. "Although Saradananda did not begin his work until more than twenty years after Ramakrishna's death, there is no doubt of its authenticity" (Christopher Isherwood, op.cit., p.2). His biography "covers all of Ramakrishna's life except its last few months" (ibid.). Mahendra Nath Gupta "first met Ramakrishna in 1882 and thereafter visited him regularly during the remaining four years of Ramakrishna's life. After each visit, M. noted down everything which had been said and done in his presence by Ramakrishna and those who were with him. The result is a very big book, which is probably close to totally accurate reporting" (ibid., pp.1-2). Also see Aldous Huxley's Foreword to Swami Nikhilananda (tr.), *The Gospel of Sri Ramakrishna* (Madras: Sri Ramakrishna Math, 1964), p.v.
9. See Eliot Deutsch and J.A.B. van Buitenen, op.cit., p. 123.
10. T.M.P. Mahadevan, op.cit., pp. 142-143. Spelling modified.
11. M. Hiriyanna, *Outlines of Indian Philosophy* (London: George Allen & Unwin Ltd., 1932) p.381. Spelling modified.
12. Ibid., p.381, fn.2. Spelling modified.
13. See T.M.P. Mahadevan, *The Hymns of Sankara* (Madras; Ganash & Co., 1970).
14. S. Radhakrishnan, op.cit., p.450.
15. See Eliot Deutsch and J.A.B. van Buitenen, op.cit., pp. 123-124, etc.
16. See S. Ramakrishna, op.cit., p.450; *Encyclopaedia Britannica* Vol.11 (Chicago: William Benton, 1961) p. 577.
17. One may indeed add *Karma* here too (see M. Hiriyanna, op.cit., p.381). This would constitute an important point of difference between Shankara and Ramakrishna, for the latter may be called a *jnanin* and a *bhakta*, but he can hardly be called a *karmayogin*. That description must be reserved for his disciple Swami Vivekananda. In that sense Shankara represents a two-in-one of Ramakrishna and Vivekananda.
18. Claude Alan Stark, *God of All* (Cape Cod, Mass.: Claude Starke Inc., 1974) Chapters II, III and IV.
19. Eliot Deutsch and J.A.B. van Buitenen, op. cit., p.122.
20. See Claude Alan Stark, op.cit., pp. 23-26.
21. Swami Saradananda, *Sri Ramakrishna: The Great Master* (Madras; Sri Ramakrishna Math, 1952), p.251.

22. Ibid.
23. Ibid., p.359.
24. Ibid., p.387.
25. Swami Nikhilananda, *The Gospel of Sri Ramakrishna* (New York: Ramakrishna-Vivekananda Center, 1942) pp. 103-4. Emphasis added.
26. M. Hiriyanna, op.cit., p.317. Emphasis supplied.
27. See A.C. Bouquet, *Hinduism* (New York: Hutchinson's University Library, 1948) p.97); Wm. Theodore de Bary, ed., *Sources of Indian Tradition* Vol.1 (New York: Columbia University Press, 1958) p. 326, etc.
28. A.C. Bouquet , op. cit., p.97.

6

WAS RAMAKRISHNA AN *ADVAITIN*?

In several modern presentations of Ramakrishna Paramahamsa, there is a tendency,[1] sometimes incipient, sometimes manifest, to depict Ramakrishna as an *Advaitin* [2]. This raises the question: can Ramakrishna be called an *Advaitin?* Was he an *Advaitin?* Before this question can be answered, a prior question must be faced: who is an *Advaitin?* What does it mean to be an *Advaitin?*

In the first part of this essay an attempt will be made to state what it means to be an *Advaitin;* in the second part, an effort will be made to determine whether Ramakrishna was one.

I

It would appear from a survey of the literature of *Advaita Vedanta*[3] that, although like many a school of philosophy, it does not represent an undifferentiated homogeneous whole[4] and in fact contains two major sub-schools[5]—the *Vivarana*[6] and the *Bhamati*[7]—yet there does exist a "common core of Advaita thinking"[8] which, with "some of the specific teachings which were developed after Shankara",[9] can be summarised in its metaphysical, metapsychological, epistemological and axiological dimensions.[10] Out of these "number of basic principles" [11] which give to *Advaita Vedanta* "its distinctive nature and quality",[12] the following would appear relevant in the present context.

(1) "*Brahman* is the sole Reality; *Brahman* is Real, and, in essence, is without quality or distinction. *Brahman* stands for undifferentiated being, for pure unqualified consciousness. '*Brahman*' means *nirguna Brahman*, qualityless reality" [13] (as distinguished from *saguna Brahman).*

(2) "The world of duality, of multiplicity, of change, and process is less than 'real'. It is the product of, and is constituted by, a creative illusion, *maya. Maya* is beginningless (*anadi*) and indescribable (*anirvacaniya*) in terms of being and non-being. From the standpoint of the subject, *maya is avidya* —ignorance. It is the power of concealing reality (*avarana-sakti*) and of misrepresenting or distorting it (*viksepa-sakti*)".[14]

(3) The Self, *Atman*, is one with *Brahman.*[15]

(4) Knowledge (alone) is the way to *Moksa*. Other paths, such as *Karma or Bhakti*, are "helpful auxiliaries", but *Moksa* is not possible without *Jnana.*[16]

(5) *Moksa* may be realized during one's lifetime. [17]

In a nutshell, to be an *Advaitin* means to maintain that *nirguna Brahman* alone is the ultimate Reality; all else is less than real, is *maya;* the *Atman* is identical with *Brahman; Moksa* is attained only through *jnana* and one can be *jivanmukta*, that is, one can attain that *jnana* within a lifetime. Now where did Ramakrishna stand with regard to these doctrines?

II

In order to analyze Ramakrishna's position on the key issues, it is helpful to frame some pointed questions.

(1) Did Ramakrishna regard *nirguna Brahman* as the sole reality? The answer is: no. For Ramakrishna is on record as having declared:

> That which is Absolute (*Nirguna*), impersonal beyond all attributes, is also the same as the personal God who is with all attributes and blessed qualities. The absolute *Brahman* again is inseparable from the divine energy (*Sakti*)...
>
> In reality there is no distinction between *"Brahman"* or the impersonal Absolute, and *"Sakti"* the divine Mother. The *Brahman* and the *Sakti* are one just as fire and its burning power are one. As by the word fire, we understand its power of burning, so by the latter we know that it is the same as fire. By realising the one, both are realized ...[18]
>
> ... He who knows the *Sakti*, or the personal aspect of the absolute Being, knows the impersonal *Brahman*. Again he who has realized the absolute Noumenon has also realized the phenomenon. He who has realized *Brahman*, has also realized the personal God or divine Mother *(Sakti)*.[19]

From the point of *Advaita Vedanta*, Ramakrishna is saying something quite revolutionary. He is asserting the identity of *nirguna* and *saguna Brahman:* thus Ramakrishna does not accept *nirguna Brahman* as the sole reality. He differentiates between two terms *jnani and vijnani*. The former is one who has realized *nirguna Brahman;* the latter, the *vijnani*, is one who "sees that the Reality which is *nirguna*, without attributes, is also *saguna*, with attributes." [20]And Ramakrishna thinks more highly of the *vijnani* than the *jnani*.[21]

However, Ramakrishna does seem to say that the experience of *nirguna Brahman* represents the ultimate in spiritual experience. All the metaphors or similies he uses in this connection, invariably seem to represent the experience of *nirvikalpa samadhi* (in which the *Atman—Brahman*—and by *Brahman* here is meant *nirguna Brahman*—identity is realized and places that experience on top, sometimes quite literally. Thus just before asserting that "Shankara's non-dualistic explanation of *Vedanta* is true, and so is the Qualified Non-dualistic interpretation of Ramanuja" he says "I cannot utter a word unless I *come down* at least two steps from the plane of *samadhi*."[22] Elsewhere he says, "The *jnani* gives up his identification with worldly things, discriminating, 'Not this, not this'. Only then can he realize *Brahman*. It is like reaching the roof of a house by leaving the steps behind, one by one."[23] Thus again the non-dual experience is identified with the very top, the ultimate in spiritual experience.

In other words, though Ramakrishna is one with *Advaita Vedanta* in accepting the realization of *nirguna Brahman* as the summit of religious experience, he does not regard it as the final religious experience. For him the religious experience of coming back to the world to realize the identity of *saguna* and *nirguna Brahman* is a desirable next step. Besides, he also parts company with *Advaita Vedanta* in not regarding the *nirguna Brahman* alone as real. Thus while he accepts the ultimacy of the experience of *nirguna Brahman*, he does not accept its sole reality as *Advaita* does. One might say that in a sense, he accepts *Advaitic* eschatology but not its ontology.

(2) Does Ramakrishna equate *Maya* with *Avidya*?

The answer is no. For Ramakrishna *"Maya or Prakriti* consists of *Vidya and Avidya. Vidya* is the energy that leads Godward... But *Avidya* leads to worldliness."[24] However, Ramakrishna concurs with *Advaita Vedanta* in asserting that *Brahman* is beyond *maya*. Indeed, this, the *vijnani* too must realize (whom he considers a cut above the mere *jnani* of the *Advaita* type). Thus he says: "A Vijnani, ... realizes the absolute *Brahman* as the unchangeable Reality of the universe, firm and immutable like Mount *Sumeru*.It is beyond all activity of *Maya*."[25] The *jnani*, however, stops at that; the *jnani* says "This world is a framework of illusion,"[26] but because *Brahman* alone has become everything, therefore to the *Vijnani*, this world is a 'mansion of mirth,' [27] not a framework of illusion.

(3) Does Ramakrishna regard the *Atman=Brahman (nirguna)*?

Ramakrishna accepts this equation: "He who is *Brahman* is verily *Atman*"[28]—but also extends it: "and again, He is the *Bhagavan*".[29] This extension is not consistent with the doctrines of *Advaita Vedanta*.

(4) Does Ramakrishna claim that *jnana* leads to *Moksa*? Yes, but is not *jnana* alone that leads to *Moksa*." There are various paths which lead to the realization of the absolute Brahman. The path of *Jnani* is as good as that of *Bhakta*. JnanaYoga is true; so is BhaktiYoga. There is another path of Bhakti mixed with *Jnana* which is equally true."[30]

(5) Does Ramakrishna believe that *Moksa* can be attained within a lifetime? Yes; Ramakrishna says that "a man becomes liberated even in this life."[31]

In other words Ramakrishna shares the eschatology of *Advaita Vedanta* but not its ontology; he shares the belief in the possibility of premortem salvation in the soteriology of *Advaita Vedanta* but rejects its *Jnanayogic* exclusivism.

III

How then is one to answer the question: was Ramakrishna an *Advaitin*? The proper way of answering the question seems to be to say that in some respects he was an *Advaitin* and in some respects he was not. He was an *Advaitin* to the extent that he regarded the experience of the formless *Brahman* as the ultimate religious experience; regarded *nirguna Brahman* as beyond *Maya;* accepted the identity of *Atman* with *Brahman;* regarded *Moksa* as attainable through *Jnana* and within a lifetime. He was not an *Advaitin* to the extent that he regarded *Saguna Brahman* as no less real than *nirguna* Brahman; did not equate *Maya* only with *Avidya*; and regarded *Moksa* as attainable through *bhakti*—as a matter of fact he seemed to recommend *bhakti* over *jnana*. In view of the fact that he used the term *vijnana* in a special sense we may—in the tradition of such descriptions as *kevaladvaita, visistadvaita, dvaita, suddhadvaita, dvait-advaita, sivadvaita,* etc.—describe him as an exponent of *Vijnanadvaita*![32]

References

1. Harold W. French, *The Swan's Wide Waters* (London: Kennikat Press, 1974) pp. 30-31, wherein it is stated that after "the realization of *Brahman* without name and form," "Ramakrishna became convinced through this experience that all *sadhanas* took the aspirant towards the non-dual plane". T.M.P. Mahadevan says that "with the realization of the non - dual Spirit, Sri Ramakrishna's life had become full. The peace that was now his, knew no disturbance because there was no duality in it" (*Outlines of Hinduism* [Bombay: Chetana Ltd., 1956] p.222). It has also been said of his biographer, Swami Saradananda, that he "insists that Sri Ramakrishna is an *Advaitin*. He says that Sri Ramakrishna's long years of *sadhana* reached the climax in *nirvikalpa samadhi* under the guidance of Tota Puri. On the basis of this experience, he says, Sri Ramakrishna came to the conclusion that *Dvaita, Visistadvaita* and *Advaita* form a progression in the spiritual development of the *sadhaka*. Interpreting Ramakrishna's views on *Advaita*, he says, 'Finally, when man reaches the ultimate limit of spiritual progress with the help of Sadhana, he experiences the *Nirguna* nature of the divine Mother and remains in oneness with Her... Swami Saradananda says that Sri Ramakrishna told his disciples that non-dualism is the ultimate truth to be realized'" (Nalini Devdas, *Sri Ramakrishna* [Bangalore: The Christian Institute for the Study of religion and Society, 1966] pp. 57-58. Emphasis added).
2. This tendency may owe something to the fact that Swami Vivekananda, Ramakrishna's best-known disciple, who was a founder-member of the Ramakrishna *matha* and mission, was *Advaitically* inclined (see R.C. Zaehner, *Hinduism* [Oxford University Press, 1962] p.223; also see R.C. Zaehner, ed., *The Concise Encyclopaedia of Living Faiths* [Boston: Beacon Press, 1959] pp.257-258, 259). Thus "Mayavati, where the Math was relocated in 1899, was conceived as an *Advaita* shrine, wholly devoted to the worship of the absolute without benefit of rituals or symbols" (Harold W. French, op.cit., p. 71). Vivekananda explicitly states: "Devotion as taught by Narada, he used to preach to the masses, those who were incapable of any higher training. He was used generally to teach dualism. As a rule, he never taught *Advaitism*. But he taught it to me. I had been a dualist before" (*The Complete Works of Swami Vivekananda* Vol.VII [Calcutta: Advaita Ashrama, 1969] p. 412). There also seems to be a hint in Vivekananda's discussion "on Sri Ramakrishna and his views" (op.cit., pp. 411-412) that the doctrine

Ramakrishna taught varied with whether someone was in the "inner" or "outer" circle. In any case, it must have varied according to the person involved as he "used to see that different people required different treatment" (ibid., p.412). This tendency to depict Ramakrishna as an *Advaitin* has been questioned for its accuracy (see R.C. Zaehner, op. cit., p. 218; Agehananda Bharati, *The Ochre Robe*, London: George Allen and Unwin Ltd., 1961, p.220, etc.). Thus another school of scholars prefers to call Ramakrishna a *bhakta*. According to R.C. Zaehner, "Ramakrishna was by nature a *bhakta* [sic], not a *jnanin*, a worshipper of God rather than an introspective, seeking the One within himself. Despite his Advaitin teacher, however, his trances did not reveal a static monad, one without a second, they showed rather that 'God had become all things'. Intellectually he may deny this on occasion, but Ramakrishna set up no store by the intellect. His followers, however, have developed philosophically into a purely monistic direction, denying all real existence to all except the One. This, however, would seem to falsify Ramakrishna's actual experience, for, for him, the infinite and the finite are not distinct but the finite is wholly suffused with the infinite; the world is God and God *is* the world" (op.cit., p.218). The remarks need to be carefully analyzed: (1) One should realize that when one calls Ramakrishna a *bhakta*, he is a *bhakta* of a special type—one who has had the non-dual experience which falls outside the pale of orthodox Bhaktic soteriology (see Krishna Datta Bharadwaj, *The Philosophy of Ramanuja* [New Delhi: Sir Shankar Lall Charitable Trust Society, 1958] pp. 215-225). (2) Ramakrishna's actual experience is not limited to the immanence of God in the world but also includes an experience of His transcendence. "Mark this", he said, "form and formlessness belong to the same Reality" (Nikhilananda, infra., p. 370. Also see p. 733). (3) One should note that "Sri Ramakrishna's teaching of *bhakti* differs from the *Advaita* of Shankara as well as from the *Visistadvaita* of Ramanuja. Shankara's teaching is that *bhakti* is valuable at a particular stage of spiritual development in order to remove the mind from the distracting forces and to make it one-pointed in its devotion. But unless *bhakti* is superseded by *jnana*, the knowledge that *Brahman* is the non-dual Reality, liberation is not possible. Sri Ramakrishna agrees with Shankara that unless all trace of distinctions is removed in the awareness of the non-dual Reality, liberation is not attained, but unlike Shankara, he is convinced that *bhakti* by itself can bring this awareness of non-duality and lead to liberation. According to Ramanuja *bhakti yoga* is the culmination of the spiritual discipline, but distinctions remain in the Goal—the finite self knows itself as a distinct entity within the Whole, the Absolute. According to Ramakrishna, those *bhaktas* who keep the trace of distinction—the feeling of 'I' and 'Thou', *bhakta* and *Bhagvan*—deliberately renounce liberation and stay within the wheel of *samsara* to enjoy the love of God. But through love they can reach the non - difference and transcend the cycle of rebirths" (Nalini Devadas, op.cit., pp. 17-18). Once Ramakrishna was asked: "Does this 'I' of the devotee never disappear altogether?" And he replied, "Yes, it disappears at times. Then one attains the knowledge of *Brahman* and goes into *samadhi*. I too lose it, but not all the time. In the music scale, there are seven notes: sa, re, ga, ma, pa, dha, and ni. But one cannot keep one's voice on 'ni' a long time.

One must bring it down again to the lower notes. I pray to the divine Mother, 'O Mother, do not give me *Brahmajnana*'". (Swami Nikhilananda, op. cit., p. 481. Emphasis added).

3. For general bibliography see Karl H. Potter, *Bibliography of Inaian Philosophies* (Delhi: Motilal Banarsidass, 1970) pp. 538-567; but also see S. Radhakrishnan, *The Vedanta According to Samkara and Ramanuja* (London: George Allen & Unwin 1928); *Indian Philosophy*, Vol.II (London: George Allen & Unwin, 1971) Chapter VIII; Surendranath Dasgupta, *A History of Indian Philosophy* (Cambridge University Press, 1963) Vol.I, Chapter X, and Vol.II, Chapter XI; etc.; but especially see Eliot Deutsch and J.A.B. van Buitenen, *A Source Book of Advaita Vedanta* (Honolulu: The University Press of Hawaii, 1971) passim.
4. ibid.; Also see T.M.P. Mahadevan, *The Philosophy of Advaita with special reference to Bharatitirtha Vidyaranya* (Madras: Ganesh & Co., 1957) etc.
5. T.M.P. Mahadevan, *Outlines of Hinduism*, pp.144-145.
6. Eliot Deutsch and J.A.B. van Buitenen, op.cit., p. 242.
7. S.S. Hasurkara, *Vacaspati Misra on Advaita Vedanta* (Darbhanga: Mithila Institute, 1948) passim.
8. Eliot Deutsch and J.A.B. van Buitenen, op.cit., p. 308. Emphasis added.
9. ibid, Diacritics omitted.
10. ibid., pp. 308-312.
11. ibid., p. 308.
12. ibid.
13. ibid. Diacritics omitted; emphasis, except for *nirguna*, supplied.
14. ibid., pp. 308-309. Diacritics omitted.
15. ibid., p. 310.
16. ibid., p.312.
17. ibid.
18. At this stage Ramakrishna uses a series of analogies to illustrate his point: They are one just as milk and its whiteness are one. We cannot conceive of milk without its whiteness.

    They are one just as a gem and its lustre are one. We cannot conceive of a gem without its lustre.

    They are one just as the serpent and its sinuous motion are one. We cannot conceive of a serpent without its serpentine movements.
19. Swami Abhedananda, *Memoirs of Ramakrishna* (Calcutta: Ramakrishna Vedanta Math, 1957) pp.108-109. Emphasis added.
20. Swami Nikhilananda, *The Gospel of Sri Ramakrishna* (New York: Ramakrishna-Vivekananda Center, 1942) p.104.
21. Swami Nikhilananda, op.cit., pp.103-104, 477-478.
22. ibid., p.733, emphasis added.
23. Swami Nikhilananda, op.cit., p.103. Emphasis added.
24. Swami Abhedananda, op.cit., p.93. Emphasis added.
25. ibid., p.93. Emphasis added.
26. Swami Nikhilananda, op.cit., p.523.
27. ibid., p.478. R.C. Zaehner quotes him thus on *Maya:* "The universe is created by

the great *maya* of God. The great *maya* contains both the *maya* of knowledge and the *maya* of ignorance. Through the help of the *maya* of knowledge, one cultivates such virtues as the taste for holy company, knowledge, devotion, love and renunciation. The *maya* of ignorance consists of the five elements and the objects of the five senses... These make one forget God" (*Hinduism*, op.cit., p.218. Diacritics omitted).

28. Swami Nikhilananda, op.cit., p.134.
29. ibid.
30. Swami Abhedananda, op.cit. p.92. Emphasis added.
31. Swami Nikhilananda, op.cit., p.159; also see p. 251. Also see Nalini Devadas, op. cit., p. 21.
32. P.N. Srinivasachari, however, has suggested the term *Anubhavadvaita* to designate Ramakrishna's position (see his *Aspects of Advaita*, [Mylapore: Sri Krishna Liberar, 1949] p.84, quoted by Nalini Devdas, op.cit., p.62, as *Anbhavadvaita* [sic]).

7

WAS RAMAKRISHNA A *VISISTADVAITIN?*

I

Ramakrishna has often been looked upon as a *jnanin or Advaitin*[1]. Such a representation of Ramakrishna has met with increasing criticism at the hands of those who choose to regard him as a *bhakta*[2] or at least a *jnanin-cum-bhakta*[3]. One is thus naturally led to ask: if it be granted that Ramakrishna was not an *Advaitin or a jnanin* then was he, in fact, a *bhakta?*[4] *Bhakta* being a somewhat loose and floating term unless pinned down to a particular philosophical school (like the term *jnanin* unless pinned down, say, to *Advaita)* one is led to ask: was Ramakrishna a *Visistadvaitin*?

II

To begin with, there are aspects of the teachings of Ramakrishna which connect neatly with certain doctrines of *Visistadvaita.*

(1) One is tempted to say, for instance, that Ramakrishna was a *Visistadvaitin* on the basis of the following excerpt:

> Master:... Shankara's Non-dualistic explanation is true, and so is the Qualified Non-dualistic interpretation of Ramanuja.
> Narendra: What is Qualified Non-dualism?
> Master: It is the theory of Ramanuja. According to this theory, *Brahman* or the Absolute, is qualified by the universe and its living beings. There three—*Brahman*, the world, and living beings—together constitute One. Take for instance a *bel* fruit. A man wanted

to know the weight of the fruit. He separated the shell, the flesh, and the seeds. But can a man get the weight by weighing only the flesh? He must weigh flesh, shell and seeds together. At first it appears that the real thing in the fruit is the flesh, and not its seeds or shell. Then by reasoning you find that the shell, seeds, and flesh all belong to the fruit; the shell and the seeds belong to the same thing that the flesh belongs to. Likewise, in spiritual discrimination one must first reason, following the method of 'Not this, not this.' God is not the universe; God is not the living beings; *Brahman* alone is real and all else is unreal. Then one realizes, as the bel fruit, that the Reality from which we derive the notion of *Brahman* is the very Reality that evolves the idea of living beings and the universe. The *Nitya* and the *Lila* are the two aspects of one and the same Reality; therefore, according to Ramanuja, *Brahman* is qualified by the universe and the living beings. This is the theory of Qualified Non-dualism.[5]

(2) Another point of possible agreement between Ramakrishna and *Visistadvaita* is the doctrine of *avatara*. As Geoffrey Parrinder has pointed out: "The teaching of Ramakrishna and his disciples is remarkable in that while it accepts Shankara's monism, it also teaches the manifestation of the impersonal *Brahma*n in the world in all the Avatars."[6] Though this statement underestimates to a certain extent the acceptance of the doctrine of *avatara* by Shankara[7] and his school, it does highlight the fact that the doctrine of avataras is theologically more developed in *Visistadvaita* and to that extent Ramakrishna's position is closer to that of Ramanuja than of Shankara.[8] Ramakrishna identified himself as an *avatara* as well.[9]

(3) There is much greater philosophical acceptance of the role of the goddess in both Ramakrishna's spiritual experiences and *Visistadvaitic* theology.[10]

(4) The open advocacy of *Bhakti* by both *Visitadvaita Vedanta* and Ramakrishna provides the strongest case for regarding Ramakrishna as belonging to that school.

III

On closer inspection, however, all these comparisons seem to fall short in some ways.

(1) In explaining the doctrine of Ramanuja to Narendra, Ramakrishna says clearly: *"in spiritual discrimination one must first reason, following the method of 'Not this, Not this.'"*[11] This process Ramakrishna clearly associates with the path of *jnana*.[12] In other words, Ramakrishna opts for 'Qualified Non-dualism' *after* the Unqualified Non-dualism has been experienced.

(2) On the question of *avatara*, Ramakrishna did make the claims outlined earlier. However, although Ramakrishna declared: "He who was Rama and he who was Krishna is now Ramakrishna in this body lying here", he also added, "Not in your Vedantic sense" (That is to say, not merely in the sense of identity with the Absolute but in the sense of Incarnation).[13]

The statement, however, is capable of an opposite interpretation that this *Avatara*-talk does not apply to *Advaita Vedanta* in its supernal version!

(3) The important role of the goddess perhaps reflects *Tantrika* rather than *Visistadvaitic* influence.[14]

(4) The notion of *Bhakti* developed by Ramakrishna is *not* identical in all ways with the one found in Ramanuja. "According to Ramanuja, *bhakti yoga* is the culmination of the spiritual discipline, but distinctions remain in the Goal—the finite self knows itself as a distinct entity within the whole, the Absolute. According to Ramakrishna, those *bhaktas* who keep the trace of distinction, the feeling of 'I' and 'Thou', *bhakta* and *Bhagvan*, deliberately renounce liberation and stay within the wheel of *samsara* to enjoy the love of God. But through love they can reach the non-difference and transcend the cycle of rebirths."[15]

V

Ramakrishna, moreover, cannot be called a *Visistadvaitin* because:

(1) he declares that "Shankara's Non-dualistic explanation of Vedanta is true", which *Visistadvaita* criticizes as untrue;

(2) the metaphors used by Ramakrishna place the non-dual experience on the mountain top while the typical *Visistadvaitic* metaphor for a Shankarite is of one who resorts to one of the caves in the mountain and never makes it to the top; [17]

(3) for Ramakrishna salvation is attainable in this life, for Ramanuja it is a post-mortem state;[18]

(4) there are also terminological differences. For Ramanuja "the difference between a *jijnasu* and a *jnani* is that the former aims at isolation (*kaivalya*), while the latter does at the realization of God."[19] Ramakrishna draws a distinction between a *Jnani* and a *Vijnani* and uses the word *Jnani* in a sense different from Ramanuja's;

(5) Ramakrishna says: *"First of all,* reach the indivisible *Satchidananda,* and *then,* coming down, look at the universe. You will then find that everything is Its manifestation."[20] Note that the non-dual experience is recommended first. It is not so in *Visistadvaita*, as noted earlier.

How then are we to describe Ramakrishna, if we must? It seems to this writer that she should be described as a *Vijnanadvaitin*. He may perhaps be called an *Advaitin* in his eschatology and perhaps also in his soteriology, but it is difficult to call him *a Kevaladvaitin;* he emphasised *Bhakti* but it is hard to call him a *Visistadvaitin*. The same holds true of other appellations. For Ramakrishna, the ultimate state is represented by a post-*advaitic* mode of existence in this world in the manner of a *vijnanin,* which he has described at various points.[21] Hence the appropriateness of the description. The following statements about the state of *vijnana* and about the *vijnanin* are particularly helpful in understanding what he had in mind:

> First of all, you must discriminate, following the method of "Neti, neti": "He is not the five element, nor the sense-organs, nor the mind, nor the intelligence, nor the ego. He is beyond all these cosmic principles". You want to climb to the roof; then you must eliminate and leave behind all the steps, one by one. The steps are by no means the roof.... But after reaching the roof you find that the steps are made of the same materials—brick, lime, and brick-dust—as the roof. It is the Supreme *Brahman* that has become the universe and its living beings and the twenty-four cosmic principles. That which is Atman has become the five elements. You may ask why the earth is so hard, if it has come out of *Atman*? All is possible through the will of God ....
>
> After attaining *vijnana* you can live in the world as well. Then one clearly realizes that God Himself has become the universe and all living beings, that He is not outside the world.
>
> The *jiva* at first remains in a state of ignorance. He is not conscious of God, but of the multiplicity. He sees many things around him.

On attaining knowledge he becomes conscious that God dwells in all beings. Suppose a man has a thorn in the side of his foot. He gets another thorn and takes out the first one. In other words he removes the thorn of *ajnana*, ignorance, by means of the thorn of *jnana*, knowledge. But on attaining *vijnana*, he discards both thorns, knowledge and ignorance. Then he talks intimately with God day and night. It is no mere vision of God.
He who has merely heard of milk is "ignorant". He who has seen milk has "knowledge". But he who has drunk milk and been strengthened by it has attained *vijnana*.
God keeps a little "I" in His devotee after giving him the knowledge of *Brahman*. That "I" is the "I of the devotee", the "I of the *jnani*" Through that "I" the devotee enjoys the infinite play of God. ... The *vijnani* retains the "I of the devotee", the "I of the *jnani*", in order to taste the Bliss of God and teach people.[22]

References

1. See Harold W. French, *The Swan's Wide Waters* (London: Kennikat Press, 1974) pp.30-31; T.M.P. Mahadevan, *Outlines of Hinduism* (Bombay: Chetana Ltd., 1956) p.222; Haridas Bhattacharyya, ed., *The Cultural Heritage of India* (Calcutta: The Ramakrishna Institute of Culture, 1956) p.674; V.S. Narvane, *Modern Indian Thought* (New York: Asia Publishing House, 1964) Chapter III; P.T. Raju, *The Philosophical Traditions of India* (University of Pittsburg Press, 1971) p.230; Swami Saradananda, *Sri Ramakrishna, The Great Master* (Madras: Sri Ramakrishna Math, 1952) p.704; R.C. Zaehner, ed., *The Concise Encyclopaedia of Living Faiths* (Boston: Beacon Press, 1959) p.257, and perhaps also Wm. Theodore de Bary, *Sources of Indian Tradition* Vol.II (New York: Columbia University Press, 1958) p.85. The portrayal of Ramakrishna as an *Advaitin* may also in part be due to the influence of his famous disciple Swami Vivekananda, see Amaury de Riencourt, *The Soul of India* (London: Jonathan Cape, 1961) p.243; K.M. Sen, *Hinduism* (Baltimore: Penguin Books, 1973 [first published 1961] p.110; etc.
2. See F. Max Mueller, *Ramakrishna: His Life and Sayings* (Bombay; Longmans, Green & Co, 1898) p. 94; R.C. Zaehner, *Hinduism* (Oxford University Press, 1966) p. 165; Thomas J. Hopkins, *The Hindu Religious Tradition* (Belmont, California: Dickenson Publishing Co., 1971) p. 137; S.C.F. Brandon, *A Dictionary of Comparative Religion* (New York: Charles Scribners Sons, 1970) p.529; Arthur Osborne, *Buddhism and Christianity in the Light of Hinduism* (London: Rider &

Co., 1959) p.52; Agehananda Bharati, *The Ochre Robe* (London: George Allen & Unwin, 1961) p.220, etc.

3. See John B. Noss, *Man's Religions* (New York: The Macmillan Co., 1967) p.301; Joseph Campbell, ed., Heinrich Zimmer, *Philosophies of India* (London: Routledge & Kegan Paul, 1953) Part III, Chapter V, Section 1.
4. For a consideration of this issue in the broader framework of Ramakrishna's experiences, see Swami Nikhilananda (tr.) *The Gospel of Sri Ramakrishna* (New York: Ramakrishna-Vivekananda Center, 1942); J.N. Farquhar, *Modern Religious Movements in India* (New York: The Macmillan Co., 1915) pp.188-208; Claude Alan Stark, *God of All* (Cape Cod, Mass.: Claude Stark Inc., 1974), etc.
5. Swami Nikhilananda (tr.) op.cit., pp. 733-734. Emphasis added.
6. Geoffrey Parrinder, *Avatar and Incarnation* (London: Faber and Faber, 1907) p.170. Emphasis added, spelling of Shankara modified.
7. See T.M.P Mahadevan, *The Hymns* of *Shankara* (Madras: Ganesh & Co., 1970); see also Haridas Bhattacharyya, ed., op.cit., Vol.1, p.179, fn. 26.
8. John Braisted Carman, *The Theology of Ramanuja* (Yale University Press, 1974) p. 179ff.
9. Romain Rolland, *The Life of Ramakrishna* (Calcutta: Advaita Ashrama, 1960) p. 273, fn.2. The following comment by Max Mueller is also of interest: "But every now and then strong fits of God-consciousness came upon him. He then became changed into a different being altogether. He then spoke of himself as being able to do and know everything. He spoke as if he had the power of giving anything to anybody. He would speak of himself as the same soul that had been born before as Rama, as Krishna, as Jesus, or as Buddha, born again as Ramakrishna .... He said that he was free from all eternity, and the practices and the struggles after religion meant to show the people the way to salvation. He had done all of them alone. He would say he was a *nityamukta*, or eternally free, and an incarnation of God Himself" (op.cit., p 58). The idea of *nityamuktas* may also be compared to that of *nityasuris* in *Visistadvaita* (John Braisted Carman, op.cit., pp. 168-169. Yet here too one may recall Ramakrishna's own statement that "From the standpoint of discrimination this 'ego of the devotee' keeps him a little away from God" (Swami Nikhilananda, op.cit., p.103).
10. See Haridas Bhattacharyya, ed., op.cit., Vol.IV, p. 663; John Braisted Carman, op.cit., Chapter XVIII.
11. Swami Nikhilananda, op.cit., p.733.
12. Joseph Campbell, ed., op.cit., p. 562.
13. Romain Rolland, op.cit., p.273, fn.2.
14. Joseph Campbell, ed., op.cit., pp.560-580.
15. Nalini Devdas, *Sri Ramakrishna* (Bangalore: The Christian Institute for the study of Religion and Society, 1966) pp. 17-18.
16. Swami Nikhilananda, op.cit., p.733
17. Personal communication, Walter G. Neevel, Jr.
18. See T.M.P. Mahadevan, op.cit., p.155.
19. Krishna Datta Bharadwaj, *The Philosophy of Ramanuja* (New Delhi: Sir Shankar Lall Charitable Trust Society, 1958) p.217.
20. Swami Nikhilananda, op.cit., p.395. Emphasis added.
21. See Swami Nikhilananda, op.cit., pp. 103-104, 404-405, 417-418, 444-479, etc.
22. Swami Nikhilananda, op.cit., pp. 417, 479, 404.

8

RAMAKRISHNA PARAMAHAMSA AND MAHATMA GANDHI: A STUDY IN CONTRASTIVE THEISMS

Ramakrishna Paramahamsa and Mahatma Gandhi (1896-1948 A.D) are two great figures of modern Hinduism.[1] They shared a common characteristic (not only between themselves but also with several other leaders of the Hindu renaissance of the last two centuries): they were theists.[2] This is not to say that they had no interest in *Advaita*,[3] but it can hardly be gainsaid that they were primarily theists.

Yet, although both of them were theists, their conception of God and the nature of their devotion to Him in some ways represents two different types of theism. This essay is an attempt to contrast the different types of theism they represent.

I

Ramakrishna was deeply committed to God. It is well known how the would-be Swami Vivekananda asked him point-blank: "Sir, have you seen God?" It is equally well-known that Ramakrishna's answer was a direct: "Yes, I have seen Him more vividly than I see you, and you can see Him too."[4] Indeed, if the accounts of Ramakrishna's experiences are to be believed—and no sound reason[5] has yet been adduced to doubt them—Ramakrishna had not one but several experiences of God. His first major experience of God consisted in his vision of the Divine Mother.[6] This experience has already been described in Chapter I.[7]

In this vision of Kali, Ramakrishna only had a vision of her. What is more significant from our point of view is that later on, he attests to the actual existence of the goddess as a "palpable" *person*. This fact comes out dramatically in the following account:

> Previously ... he saw that there appeared a wonderful living presence in the stone image before him. Now he did not see that image at all when he entered the temple; but saw instead, standing there the living Mother Herself, all consciousness... The Master said, "I put the palm of my hand near her nostrils and felt that Mother was actually breathing. I observed very closely, but I could never see the shadow of the Mother's divine person on the temple wall in the light of the lamp at night. I heard from my room that Mother, merry like a little girl, was going upstairs, her anklets making jingling sounds. I came up to test it and found that She, with Her hair dishevelled, was actually standing on the verandah of the first floor of the temple and was now viewing Calcutta, now the Ganga."[8]

The important point to note here is that the goddess was for Ramakrishna an *actual living perceptible person*. The same is true of his experience of God. He met God as a person:

> Yes, God talked to me.... I saw Him emerging from the Ganga and walking towards the banyan tree. We both laughed a good deal. He cracked my fingers in fun, and then He talked and talked...[9]

Ramakrishna even testifies here to the divine garrulousness of God as a person!

Thus one aspect of Ramakrishna's theism to be borne in mind is that God to him was also *a person*, apart from whatever other things He was. Another feature of Ramakrishna's theism is that it seems to be saturated with a sense of blissfulness.

Many of the experiences of God which Ramakrishna had were characterized by the experience of bliss. While narrating his subsequent experiences of the Divine Mother, he described her form as "consisting of pure consciousness and *bliss*."[10] His vision of Sita culminated in the following experience:

> I was about to fall at her feet crying, "Mother", when she entered into my body with the significant remark that the smile on her lips

> she bequeathed to me! I fell unconscious on the ground, overpowered with emotion.[11]

Similarly, in his experience of the child Rama or *Ramalala* "his mind was ... filled with ... that wonderful love and attraction which a mother feels towards her young child".[12]

Ramakrishna's experiences in the *madhura bhava* also ended in moments of ecstasy. In the practice of this *bhava*, Ramakrishna assumed the role of Radha and

> Sri Krishna revealed Himself to him in His exquisitely graceful form, and fulfilled the longing of his soul. Then He merged Himself in the person of Sri Ramakrishna, who remained for three or four months in a state of divine felicity.[13]

Such instances could be multiplied. They all point to the fact that one remarkable feature of Ramakrishna's experience of God was that it was ecstatic. This is not to suggest that no pain was involved. Even in his very first experience of Kali, Ramakrishna testifies to the excruciating pain he felt prior to the experience. Nevertheless, it seems reasonable to suggest that in the actual experience of God he felt ecstasy. This is confirmed by the ecstatic trances or *bhavas* he was prone to.

II

Mahatma Gandhi was also deeply committed to God. He refers to Charles Bradlaugh's brand of atheism in his autobiography:

> And how could I help knowing something of atheism too? Every Indian knew Bradlaugh's name and his so-called atheism. I read some book about it, the name of which I forget. It had no effect on me for I had already crossed the Sahara of atheism.[14]

Mahatma Gandhi's encounter with God, however, hardly ever took the form of a vision of God and hardly ever resulted in ecstasy. Rather, Mahatma Gandhi talks of God in the context of moral crises in his life. This seems to point to a distinction in the manner in which Ramakrishna and Mahatma Gandhi encountered God. In both cases, emotional intensity is involved but in the case of Ramakrishna, emotional intensity was associated with devotional fervour, in the case of Mahatma Gandhi with moral travail. While describing his glimpses of religion from "my sixth or seventh year up to my sixteenth."[15] Mahatma Gandhi writes:

> But one thing took deep root in me — the conviction that morality is the basis of things, and that truth is the substance of all morality. Truth became my sole objective. It began to grow in magnitude every day, and my definition of it also has been ever widening.[16]

Later on Mahatma Gandhi observes that he "was confirmed in his opinion that religion and morality were synonymous".[17] If one wished to paraphrase the experiences of Ramakrishna along similar lines, one would perhaps have to say that for him religion and ecstasy, rather than morality, were synonymous.

Just as Ramakrishna experienced God in moments of ecstatic devotion, Mahatma Gandhi experienced God in moments of moral perturbation. And what he experienced in these moments was not the vision of God but the saving power of God. This saving power of God is clearly stated on the "occasion on which a woman, other than my wife, moved me to lust".[18]

> I did not then know the essence of religion or of God, and how He works in us. Only vaguely I understood that God had saved me on that occasion. On all occasions of trial, He has saved me. I know that the phrase 'God saved me' has a deeper meaning for me today, and still I feel that I have not yet grasped its entire meaning. Only richer experience can help me to a fuller understanding. But in all my trials — of a spiritual nature, as a lawyer, in conducting institutions, and in politics—I can say that God saved me.[19]

One should note, however, that although the fact of having been saved by God clearly implies a *personal relationship* with God, for Mahatma Gandhi, God was *not* a person in the sense he was for Ramakrishna, though a devotee was seen as capable of entering into a personal relationship with Him. Thus Mahatma Gandhi wrote:

> God is not a person ... God is the force. He is the essence of life. He is pure and undefiled consciousness. He is eternal. And yet, strangely enough, all are not able to derive either benefit from or shelter in the all-pervading living presence.

> Electricity is a powerful force. Not all can benefit from it. It can only be produced by following certain laws. It is a lifeless force. Man can utilize it if he labours hard enough to acquire the knowledge of its laws.
>
> The living force which we call God can similarly be found if we know and follow *His law* leading to the discovery of Him in us.[20]

Thus Mahatma Gandhi could not feel the breath of *Kali* or laugh with God as Ramakrishna did but he could find Him in his own heart and conscience.

III

Both Ramakrishna Paramahamsa and Mahatma Gandhi were theists; they were not only theists but also *bhaktas*. Ramakrishna was pre-eminently a devotee of Kali and Mahatma Gandhi of Rama. Both of them slightly deviated from their formal religious backgrounds in the choice of their *ista-devata*. "Ramakrishna's family deity was Raghuvir, an aspect of Ramachandra", [21] yet his devotion focused primarily on Kali. It would appear that in the *Vaishnavism* practised in Mahatma Gandhi's home, Krishna would have figured more prominently than Rama. This certainly seems to hold true for Gujarat in general. Yet Mahatma Gandhi's own devotion focused primarily on Rama. When Ramakrishna died he uttered the name of Kali.[22] When Mahatma Gandhi died he took the name of Rama[23] and both thus measured up to the supreme test of devotion in Hindu theism—taking the name of God at the time of one's death.[24]

Yet the content of the theism of Ramakrishna and Mahatma Gandhi seems to have differed: for one, God was a person and His experience essentially ecstatic, for the other, God was a spiritual force which was primarily encountered morally.

References

1. See T.M.P. Mahadevan, *Outlines of Hinduism* (Bombay: Chetana Ltd., 1960) pp. 216-231.
2. See V.S. Narvane, *Modern Indian Thought* (New Delhi: Asia Publishing House, 1964) passim.
3. See Claude Alan Stark, *God of All* (Cape Cod, Mass.: Claude Stark Inc., 1974) pp. 57-60; M.K. Gandhi, *Hindu Dharma* (Ahmedabad: Navajivan Publishing House, 1958) p.55.
4. Claude Alan Stark, op.cit., p. 124.
5. See J.N. Farquhar, *Modern Religious Movements in India* (New York: Macmillan Co., 1915) p.189, etc.; also see Alexander Lipski, "Some Methodological problems Encountered by Western scholars in the Study of Modern Indian Religious History", in *Journal of Indian History,* Vol.43 (1965) pp. 275-276.
6. Swami Saradananda, *Sri Ramakrishna, The Great Master* (Madras: Sri Ramakrishna Math, 1952) p. 71.
7. See note 24 of Chapter 1.
8. Quoted in Claude Alan Stark, op.cit., p. 29.
9. Quoted in V.S. Narvane, op. cit., pp. 76-77.
10. Swami Saradananda, op.cit., p. 145. Emphasis added.
11. ibid., p. 83.
12. ibid., p. 142.
13. Quoted in Claude Alan Stark, op.cit., p. 101.
14. Mahatma Gandhi, *Story of My Experiments with Truth* (Washington DC.: Public Affairs Press, 1948) p.92.
15. ibid., p. 47.
16. ibid., pp. 50-51.
17. ibid., p. 207.
18. ibid., p. 95.
19. ibid., p. 96. In this context it is interesting to compare the relation of Good and Evil to God as seen by Ramakrishna and Mahatma Gandhi. Ramakrishna states that "God tells the thief to go and steal, and at the same time, warns the householder against the thief". Mahatma Gandhi also says that "In a strictly scientific sense, God is at the bottom of both good and evil. He directs the assassin's dagger no less than the surgeon's knife. But for all that, good and evil are, for human purposes, from each other distinct and incompatible, being symbolical of light and darkness, God and Satan" *(All Men are Brothers,* UNESCO 1959, p. 68 [50]). Thus while Ramakrishna and Mahatma Gandhi both seem to accept the ontological *transethicism* of God, they differ on the constitutive nature of His personality and the prescriptive nature of morality. In this sense Ramakrishna's comment on stealing may be compared with Mahatma Gandhi's own repentance at having stolen (op.cit., Chapter VIII).
20. *All Men Are Brothers,* p.65(40). Emphasis added.
21. Claude Alan Stark, op.cit., p. 23.
22. Romain Rolland, *The Life of Ramakrishna* (Calcutta: Advaita Ashrama, 1960) p. 274.
23. J.B. Kripalani, *Gandhi: His Life and Thought* (Government of India: Publications Division, 1917) p. 301.
24. *Bhagavadgita* VIII. 5.

9

RAMAKRISHNA AND PARAPSYCHOLOGICAL RESEARCH

I

Ramakrishna has had numerous biographers.[1] Out of them, Romain Rolland, seems to have been more sensitive to those aspects of his life which might provide useful areas or data for further scientific investigation. This essay is an effort to consolidate some of the suggestions made along these lines which lie scattered in his book[2]. Other authors also refer to his psychic powers[3] but do not seem to single them out with the same definiteness and interest as he does.

II

One such feature seems to be the relation of the colour of the breast and the stage of one's yogic advancement. Romain Rolland cites a "curious detail" of a meeting between Ramakrishna and Devendranath Tagore which "may interest our European psycho-physiologists".

> Hardly were the introductions over than Ramakrishna asked Devendranath to undress and show him his chest. Devendranath complied without showing much astonishment. The colour of the skin was scarlet, and Ramakrishna examined it. This persistent redness of the breast is a peculiar sign of the practice of certain Yoga. Ramakrishna never omitted to examine the breast of his disciples, their breathing capacity, and the soundness of their circulation before allowing or forbidding them to undertake exercises of great concentration.[4]

Romain Rolland further tell us that Ramakrishna recognized in Swami Dayananda "this characteristic redness of breast".[5]

III

A second area is suggested by the ability of Ramakrishna to induce a mental state by touch or sometimes even without it. Tactile telepathy or plain telepathy thus becomes a potential area of investigation. Romain Rolland mentions the name of Swami Shivananda as one of the disciples who experienced this and suggested that "the learned men of Europe, who are preoccupied by problems of mystic psychoanalysis, put themselves in touch with these living witnesses."[6] He quotes Swami Shivananda as saying:

> Ramakrishna had the power to raise others to the greatest heights of consciousness by transmitting to them the energy of his own spirituality. He did it either by the power of high thought or by his touch. Many of us had the privilege of being taken to higher planes of spiritual consciousness according to our capacity. It was neither hypnotism, nor a condition of deep sleep. I myself had the privilege of attaining this high spiritual consciousness three times through his touch and by his will. I still live to bear direct witness to his tremendous spiritual power.[7]

IV

Finally, another area of investigation is suggested by the ecstasies of Ramakrishna. "It is known that the stethoscopic examination of the heart and the condition of the eyes during *Samadhi* show all the symptoms of the condition of death."[8] Besides, the manner in which Ramakrishna could be brought down to lower planes of consciousness from his ecstasies is also not devoid of scientific interest. "For the interest of European science, it is to be noted that the only method of recalling Ramakrishna from his ecstatic trances was to pronounce in his ear such or such a name of the Lord, or some Mantra (form of prayer), differing according to the degree and the form of the ecstasy. The character of psychic concentration was then very marked; and it was impossible to speak of any initial physiological disorder; the spirit always remained in full control."[9]

References

1. See Max Mueller, *Ramakrishna: His Life and Sayings* (London: Longmans, Green & Co., 1898); Romain Rolland, *The Life of Ramakrishna* (Calcutta: Advaita Ashrama, 1960); Swami Saradananda, *Sri Ramakrishna, The Great Master* (Madras: Sri Ramakrishna Math, 1952); Christopher Isherwood, *Ramakrishna and His Disciples* (London: Methuen & Co., Ltd., 1965); etc.
2. Romain Rolland, op.cit., passim.
3. "He had attained to great yoga powers, but he never cared to display these marvellous powers to anybody.... But persons who went to him have found abundant proofs of his possessing such powers as thought - reading, predicting future events, seeing things at a distance, and healing a disease by simply willing. The one great power of which he had most use, and which was by far the most wonderful, was that he was able to change a man's thoughts by simply touching his body" (Max Mueller, op.cit., pp.53-54).
4. Romain Rolland, op.cit., p.150, fn.1.
5. ibid., p. 151.
6. ibid., p. 216, fn. 1.
7. ibid., also see p. 240, fn. 1.
8. Romain Rolland, op.cit., p. 263 fn. 1. Romain Rolland also points out that Dr Sarkar, his doctor, who did not share his religious views, "was present during several ecstasies and studied them from a medical point of view. A study of his notes would be of great interest for European science". The following account found in Max Mueller may also be of some interest on this point. "About this time Ramakrishna began to practise Yoga, or the physical discipline, which makes the body strong and enduring. He began by regulating his breath, and went through the eight-fold methods prescribed by Patanjali. His teachers were astonished at the short time in which he came to the realisation and attained the end of all these ascetic practices. One night, when he was practising Yoga, he was very much frightened at two strings of clotted blood coming out of his mouth. The temple services were then in the hands of one of his cousins, Haladhari, a man of great learning and purity and possessed of certain psychical powers, such as *Vak-siddhi*, power of speech. A few days before, Ramakrishna had offended him by pointing out to him certain defects on his character, so much so that his cousin cursed him and said that blood should come out of his mouth. So Ramakrishna was frightened, but a great Yogin who was living there at the time came to his help, and after inquiring into his case assured him that it was very good that the blood had come out that way. It was because he had to teach many men, and to do good to them, that he was not permitted to enter into that *Samadhi* (trance) from which nobody returns. He explained to him that when a man has attained to the perfection of this

Yoga his blood rushes to his brain, and he becomes absorbed in *Samadhi*, perceives his identity with the Supreme Self, and never returns any more to speak of his religious experiences to others. Only a few returned, namely, those who by the will of God were born to be the great teachers of mankind. In their case the blood rushes to the brain, and they feel the identity for some time, but after that the blood flows out again and they are able to teach" (Max Mueller, op.cit., pp. 45-46).

9. Romain Rolland, op.cit., p. 155.

10

WAS RAMAKRISHNA A HYPNOTIST?

I

The biographers of Ramakrishna are agreed that he had the ability of influencing a person's thoughts—and even inducing a trance—through his touch.[1] This has led some scholars to suggest that Ramakrishna produced this effect through hypnosis.[2] It is the purpose of this essay to examine whether this was so.

II

In hypnosis, the mental condition of the hypnotist and the subject remains distinct. It is true that the hypnotist influences the mental state of the subject through suggestion, but it is the mental state of the subject alone that is altered by the suggestion, not that of the hypnotist. For instance, when the hypnotist suggests to a subject that the subject is standing on his head, it is the subject who begins to imagine thus, *not* the hypnotist.

On the basis of the evidence deduced from various instances in which Ramakrishna influenced people's ideas by touch or induced a state of trance, it can be said that this independence between the hypnotist's state and that of the subject referred to earlier did not always necessarily obtain.

A few typical cases may be examined in this connection. What follows is a description of the "first instance of *samadhi* in the life of Swami Vivekananda ... communicated by a touch from Sri Ramakrishna," and is described as follows by Swami Saradananda, using Vivekananda's own words:

> No sooner had he seen me than he called me joyfully to him and made me sit on one end of the bedstead. He spoke indistinctly something to himself, looked steadfastly at me and was slowly coming toward me.... He came to me and placed his right foot on my body, when, immediately, I had a wonderful perception. I saw with my eyes open, that all the things of the room together with the walls were rapidly whirling and receding into an unknown region and my "I" - ness together with the whole universe was, as it were, going to vanish in the all-devouring great void. I was then overwhelmed with a terrible fear; I had known that the destruction of "I" - ness was death and that death was before me, very near at hand. Unable to control myself, I cried out loudly and said, "Ah! what is it you have done to me? I have my parents, you know". Giving out a hoarse laugh to hear those words of mine and touching my breast with his hand, he said, "Let it then cease now; it need not be done all at once; it will come to pass in course of time". I was amazed to see that extraordinary perception of mine vanish as quickly as it had come when he touched me that way and said those words. I came to the normal state and saw things inside and outside the room standing still as before.[3]

It is clear from this account that the alteration in Swami Vivekananda's state was *not* accompanied by a similar state on the part of Ramakrishna, who rather worked on Vivekananda. Thus the hypnotist-subject relationship seems to hold in this case and it could be at least plausibly argued in this case that hypnotism of however sublime a type may be involved.

A second experience of the type discussed above, however, "was told to Saradananda both by Vivekananda and Sri Ramakrishna," and is recounted as follows:

> The Master walked with Narendra for some time in the garden on the bank of the Ganga that day. Talking with Narendra, he came to that parlour, sat down there and went into *samadhi* shortly afterwards. Narendra sat near at hand, and was calmly observing that state of the Master, when the latter suddenly approached and touched him as before. Although he exercised great caution because of his previous experiences, Narendra became completely overwhelmed at that powerful touch. He lost external consciousness

> completely that day, unlike what had happened on the previous occasions. When he regained consciousness, after some time, he saw that the Master was passing his hand on his chest and, on seeing him come to the normal state, smiled gently and sweetly.
> Narendra did not tell us anything of the nature of the experience he had after he had lost his external consciousness. We thought he did not express it to us because it was a secret. But we realized later that it was natural for Narendra not to have remembered it, from what the Master told us one day in course of his conversation on this event. The Master said:
> 'When Narendra had lost his external consciousness, I asked him that day many questions, such as who he was, where he came from, why he came (was born), how long he would be here (in this world) and so on and so forth. Entering into the depth of his being, he gave proper answers to all these questions. These answers of his proved what I thought and saw and knew in my visions about him. It is forbidden to say all those things. But I have known from all these that on the day when he will know who he is, he will no more remain in this world; he will, with a strong power of will, immediately give up his body through yoga. Narendra is a great soul perfect in meditation'.[4]

One notes here the Master started by going into *samadhi* himself and then later suddenly approached and touched him as he had done before. In this case too, the hypnotist-subject relationship seems to hold, *but only in the later stages*. In the earlier stages, it seems, Ramakrishna had to go into *samadhi* to induce it in Vivekananda. In another instance Ramakrishna "then touched Narendra and went into *samadhi*".[5] Thereupon Narendra had his unique experience. This seems to confirm the earlier impression that, at least at some point, Ramakrishna *himself* had to enter the state he wished to induce in the subject. The experience of the other disciples of Ramakrishna seems to confirm this view.[6]

III

On the basis of these cases, Romain Rolland even suggested that while he had little curiosity of his own on this point, "let the learned men of Europe, who are preoccupied by the problems of mystical psychoanalysis, put themselves in touch with these living witnesses while there is yet time."[7]

The call probably then went unheeded and now it is perhaps too late. The conclusion which Romain Rolland drew on the basis of the data, however, is worth citing.

> For scientific men, who study psycho-physiological problems, it is noteworthy that these "touches", which provoked in the subjects concerned immediate experiences of changed conditions, were nearly always (if not always) produced when Ramakrishna was in a state of semi-consciousness or of complete hypnosis. There was therefore nothing in them analogous to calculated action of the will independent of the energies governed by it. It might almost be described as a forced descent of another into the abyss he had first descended himself.[8]

It is clear, then, that the tactile transmissions of Ramakrishna can hardly be regarded as forms of hypnosis, since the mental state of the 'hypnotist' and the 'subject' had to be in tune. It is as if the hypnotist, in order to put the subject to sleep, has to fall asleep himself! What might be said on the basis of these experiences of Ramakrishna is that if they must be analyzed within the framework of hypnosis then a state of *hyphnotism* (in which both the hypnotist and the subject share the state of hypnosis) seems to be involved.[9]

References

1. See Max Mueller, *Ramakrishna: His Life and Sayings* (London: Longmans Green & Co., 1898) pp.53-54; Romain Rolland, *The Life of Ramakrishna* (Calcutta: Advaita Ashrama, 1960) pp.216, 240; Swami Saradananda, *Sri Ramakrishna, The Great Master* (Madras: Sri Ramakrishna Math, 1952) p.768, etc.
2. See J.N. Farquhar, *Modern Religious Movements in India* (New York: The Macmillan Co., 1915) p.189. etc.
3. Quoted in Claude Alan Stark, *God of All* (Cape Cod, Mass.: Claude Stark Inc., 1974) pp.124-125.
4. ibid., pp.125-126.
5. ibid., p.128.
6. Romain Rolland, op.cit., p.216 fn.1.
7. ibid., p.216.
8. ibid., p.250 fn.1.
9. Whether this relationship is parasitical or symbiotic or involves neither of these relationships is a point which might interest future investigators.

11

RAMAKRISHNA FROM A BUDDHIST VIEWPOINT

I

Ramakrishna Paramahamsa was a Hindu who experimented with other religions as well.[1] He was quite aware of the existence of the Buddha. On one occasion he said:

> I have heard a great deal about Buddha. He is one of the ten Incarnations of God.[2]

On another occasion he challenged Narendra's assertion that the Buddha was an atheist by claiming that "He simply did not express his inner experience in words",[3] and engaged in a prolonged discussion with him on the Buddha.[4]

However, although Ramakrishna was aware of the Buddha, the way he was of, say, Christ or Muhammad, he did not consciously set out to follow the Buddhist path the way he did follow the Christian or the Muslim path. Though, when Claude Alan Stark speaks of the verification of God by Ramakrishna in various religions, he does have a chapter on the "Verification of 'God' in Buddhism",[5] the chapter really deals with "Sri Ramakrishna's *Nirvikalpa Samadhi* as *Nirvana*."[6] There is some justification for this perhaps from the Hindu point of view, for Ramakrishna claimed that: "There is not the least doubt about Lord Buddha's being an Incarnation. There is no difference between his doctrines and those of the Vedic *jnanakanda*."[7]

The Buddhists may differ on this point but let the Hindu viewpoint be adopted for the purposes of this essay and let it be assumed that like the

Buddha, Ramakrishna indeed experienced *Nirvana*. If one makes this rather enormous assumption, then it becomes possible to examine Ramakrishna from a Buddhist point of view.

It is well-known that Buddhism is not a single monolithic tradition and that it is encountered basically in the form of one of its two basic divisions: the *Theravada* and the *Mahayana*. Ramakrishna, therefore, may be examined successively in the light of first *Theravada* and then *Mahayana* Buddhism. In order to impart a clearer focus to the discussion it might be helpful to examine Ramakrishna's experiences for the light they may shed on issues within *Theravada* and *Mahayana* Buddhism. In *Theravada* Buddhism, for instance, there is some discussion as to permanency of the achievement of *Nirvana*. Similarly, in *Mahayana* Buddhism, much discussion centres on the figure of the *Bodhisattva*.

II

Let us first discuss the issue of *Nirvana* in *Theravada* Buddhism. The controversy relates to the question: can *Nirvana* once achieved be lost?

> According to the *Theravadins*, the acquisition of *Nirvana* is final and definite and can never again be lost. The *Sammitiyas*, *Vajjiputtiyas*, *Sabbatthivadins* and some *Mahasanghikas*, however, held that the *arhat* is liable to fall away. The *Saddharmapundarika* (v, 59-83) speaks of the *Nirvana* of the *arhats* as a temporary repose and distinguishes it from the final *Nirvana* of the Buddha.[8]

What light does Ramakrishna's experience shed on this controversy?

We continue with the assumption that in the state of *nirvikalpa samadhi* Ramakrishna had achieved *Nirvana*. We also assume that if *Nirvana* has been achieved, there can be no rebirth.[9]

It is here that we must face a striking fact about Ramakrishna, that he will be reborn. This we have on his own authority:

> My divine mother has also shown me that I shall have to come back again and that my next incarnation will be in the West.[10]

We also have this on the authority of his disciple Vivekananda, that "He said he would come back once more with us."[11]

It is clear, then, that if Ramakrishna's experience of *Nirvana* be accepted as authentic, then the claim made of *Nirvana* that "*The man who finds it never again loses it*, and when he dies he passes to this state forever",[12] may not be entirely true.

The question of *Bodhisattva* in *Mahayana* Buddhism may now be raised. Let us try to determine how a *Mahayana* Buddhist would classify Ramakrishna notwithstanding the obvious problems of such an enterprise.

One may begin by stating that Ramakrishna cannot be considered another Gautama Buddha for the obvious reason that there was only one Gautama Buddha, who passed into *Parinirvana*.[13] There are a few additional considerations to be kept in mind.

(1) Although Ramakrishna at least once, though not without hedging it, asserted his identity with Rama and Krishna, he does not seem to have ever done so with the Buddha on that final occasion.[14]

(2) A Buddha is not reborn, at least in *Theravada* Buddhism. But it appears that Ramakrishna will be reborn according to available accounts.[15]

(3) Ramakrishna claimed that he was an *avatara*,[16] and the same [17] was claimed for him in his own lifetime.[18] The Buddha never made such a claim,[19] though the claim was made for him later.[20]

Ramakrishna cannot be considered *the* Buddha, but can he be considered *a* Buddha? The achievement of *nirvikalpa samadhi* by Ramakrishna inclines one to this view,[21] but the fact that he "relapsed" in the sense that he will be reborn makes it hard to maintain this position.

It appears, however, that Ramakrishna could probably be called a *Bodhisattva* from a Buddhist point of view. Like a *Bodhisattva* he will be reborn and may be seen as having postponed his own *Nirvana*.[22] He is also reported to have said:

> I would suffer willingly all sorts of bodily pains and death also, a hundred thousand times, if by so doing I could bring a single soul to freedom and salvation.[23]

This is reminiscent of the *Bodhisattva's* resolve.

> "All creatures are in pain", he resolves, "all suffer from bad and hindering *karma*... so that they cannot see the Buddhas or hear the Law of Righteousness or know the Order.... All that mass of pain and evil *karma* I take in my own body.... I take upon myself the burden of sorrow; I resolve to do so; I endure it all, I do not turn

back or run away. I do not tremble.... I am not afraid... nor do I despair. Assuredly I must bear the burdens of all beings... for I have resolved to save them all. I must set them all free, I must save the whole world from the forest of birth, old age, disease and rebirth, from misfortune and sin, from the round of birth and death, from the toils of heresy.... For all beings are caught in the net of craving, encompassed by ignorance, held by the desire for existence; they are doomed to destruction, shut in a cage of doubts, always at loggerheads, one with another, always prone to see evil; they cannot find a refuge in the ocean of existence; they are all on the edge of the gulf of destruction.[24]

IV

It appears reasonable to conclude, therefore, that from a Buddhist point of view, Ramakrishna could perhaps be looked upon as (a) clarifying a debated point in *Theravada* Buddhism and (b) assuming a definite order of sanctification in *Mahayana* Buddhism.

References

1. See Christopher Isherwood, *Ramakrishna and His Disciples* (London: Metheuin & Co., 1965) passim; Claude Alan Stark, *God of All* (Cape Cod, Mass: Claude Stark Enc., 1974) passim; etc.
2. Swami Nikhilananda, *The Gospel of Sri Ramakrishna* (Madras: Sri Ramakrishna Math, 1964) pp.382, 959.
3. ibid., p.940.
4. ibid., pp.940-941.
5. Claude Alan Stark, op.cit., Chapter IV.
6. ibid., p.57. Emphasis added.
7. ibid., p.68.
8. G.P. Malalasekera, ed., *Encyclopaedia of Buddhism*, Volume II (Government of Ceylon, 1966) p.47:
9. Wm. Theodore de Bary, ed., *Sources of Indian Tradition* Vol.I (Columbia University Press, 1958) p.94.
10. Swami Abhedananda, *Memoirs of Ramakrishna* (Calcutta: Ramakrishna Vedanta math, 1957) p.340; but also see Romain Rolland, *The Life of Ramakrishna* (Calcutta: Advaita Ashrama, 1960) p.174, fn.2.

11. *The Complete Works of Swami Vivekananda* Vol.VII (Calcutta: Advaita Ashrama, 1969) p.413.
12. A.L. Basham, *The Wonder that was India* (New York: Taplinger Publishing Co., 1967) p.274, emphasis added.
13. Richard H. Robinson, *The Buddhist Religion: A Historical Introduction* (Belmont, California: Dickenson Publishing Co. Inc., 1970) p.35. This of course represents a *Theravada* understanding of the situation.
14. Romain Rolland, *The Life of Ramakrishna* (Calcutta: Advaita Ashrama, 1960) p.273, fn.2. On other occasions he did, though perhaps less emphatically and dramatically, see Geoffrey Parrinder, *Avatar and Incarnation* (London: Faber & Faber, 1970) p.107.
15. Romain Rolland, op.cit., p.174, fn.2.
16. Max Mueller, *Ramakrishna: His Life and Sayings* (London: Longmans, Green & Co., 1898) p.58.
17. Christopher Isherwood, op.cit., pp.99-101.
18. Romain Rolland, op.cit., p.173, fn.1, etc.
19. See Claude Alan Stark, op.cit., p.68, fn.38.
20. T.M.P. Mahadevan, *Outlines of Hinduism* (Bombay: Chetana Ltd., 1960) p.36.
21. See Claude Alan Stark, op.cit., Chapter IV.
22. ibid., p.190.
23. Max Mueller, op.cit., p.57.
24. Wm. Theodore de Bary, ed., op.cit., p.161. Also see p.160 where the *Bodhisattva* resolves to take on the sins of others and see Romain Rolland, op.cit., p.261, fn.2.

12

DID RAMAKRISHNA BELIEVE IN REINCARNATION?

I

Ramakrishna Paramahamsa is one of the major figures of the Hindu renaissance, who is often credited with having stemmed the tide of the Westernization of Hinduism as represented by bodies such as the *Brahmo Samaj*.[1] He is said to have done so by revalidating Hinduism in the eyes of his contemporaries through the personal verification of its verities.[2] One of the fundamental Hindu beliefs is that of *Karma*, with its logical corollary—that of rebirth or reincarnation.[3] It seems to have been hitherto unquestionably assumed, under the influence of the view that Ramakrishna vindicated traditional Hinduism, that Ramakrishna must have believed in reincarnation.

It is the purpose of this essay to suggest that while Ramakrishna can be said to have clearly believed in the doctrine of *avatara* or incarnation, he believed in the doctrine of reincarnation only with some measure of equivocation.

II

It is true that Ramakrishna often refers freely to the doctrine of reincarnation in his dialogues with his disciples and in his parables. Indeed, at one point he is asked pointedly: "Sir, what is the nature of life after death?" And Ramakrishna answers:

> Keshab Chand also asked that question. As long as a man remains ignorant, that is to say, as long as he has not realized God, so long

> will he be born. But after attaining Knowledge he will not have to come back to this earth or to any other plane of existence.
> The potter puts his pot in the sun to dry. Haven't you noticed that among them there are both baked and unbaked ones? When a cow happens to walk over them, some of the pots get broken to pieces. The broken pots that are already baked, the potter throws away, since they are of no more use to him. But the soft ones, though broken, he gathers up. He makes them into a lump and out of this forms new pots. In the same way, so long as a man has not realized God, he will have to come back to the Potter's hand, that is, he will have to be born again and again.[4]

This statement seems to imply a clear belief in the doctrine of reincarnation on the part of Ramakrishna. Elsewhere too Ramakrishna seems to acknowledge clearly the fact of rebirth. Thus he remarks: "After a process of severe struggle with one's lower nature and the assiduous practice of spiritual discipline leading to Self-knowledge, one attains the state of *Samadhi*. Then the ego with all its train vanishes. But it is very difficult to attain *Samadhi*; the ego is very persistent. That is why we are born again and again in this world".[5] Not only does he refer to rebirth in the context of Yoga, he also refers to it in the context of *Sakti-Sadhana*,[6] and in the context of *Bhakti*.[7]

III

A closer look at the pronouncements of Ramakrishna, however, seems to produce a slightly different impression. Thus once when he is asked point-blank "Sir, is there such a thing as reincarnation? Shall we be born again?", Ramakrishna says:

> Ask God about it. Pray to Him sincerely, He will tell you everything. Speak to Jadu Mallick, and he himself will tell you how many houses he has, and how many government bonds. It is not right to try to know these things at the beginning. First of all realize God; then He Himself will let you know whatever you desire.[8]

The qualified nature of this reply may be contrasted with his forthright response to the would-be Vivekananda's query: "Have you seen God?

'Yes, my son, I have seen God, just as I see you before me, only much more intensely'.[9] Why, then, this reluctance on the part of Ramakrishna to commit himself to a firm assertion of reincarnation?

The answer seems to lie in the fact that Ramakrishna's belief in reincarnation seems to have been traditional rather than personal. He believed in it because it is part and parcel of the Hindu tradition to which he belonged, and not because he had verified it personally. Indeed, at one point he seems to acknowledge this quite frankly. A *Brahmo* devotee asked him: "Sir, do you believe in the reincarnation of the soul?" and he replied:

> Yes, *they say there is something like that*. How can we understand the ways of God through our small intellects? *Many people have spoken about reincarnation; therefore I cannot disbelieve it.*[10]

In this response we find an admission of divine agnosticism (God alone knows) as it were, coupled with a clear statement that Ramakrishna himself knows it and accepts it only on the authority of tradition.

It is clear, therefore, that Ramakrishna's acceptance of reincarnation did not result from personal verification but from the acceptance of tradition.[11]

References

1. See D.S. Sarma, *Studies in the Renaissance of Hinduism* (Benaras Hindu University, 1944) p.288.
2. ibid.
3. R.C. Zaehner, ed., *The Concise Encyclopaedia of Living Faiths*, (Boston: Beacon Press, 1959) p.225.
4. Swami Nikhilananda, *The Gospel of Sri Ramakrishna* (New York: Ramakrishna-Vivekananda Centre, 1942) p.416. One may add facetiously that only the half-baked ones come back!
5. *Sayings of Sri Ramakrishna* (Madras: Sri Ramakrishna Math, 1965) p.58. Emphasis added.

6. "Two persons, it is said, together began invoking the goddess *Kali* by the terrible rite called *Savasadhana*. One invoker was frightened to insanity by the terrible visions appearing in the earlier portion of the night, while the other was favoured with the vision of the Divine Mother at the end of the night. The latter said to the Goddess, 'Mother, why did the other man become mad?' The Deity answered, 'You, too, o child, did become mad many times in previous births, and now at last you have seen me'" (ibid., p.177).
7. "That man whose hair stands on end at the mere mention of the name of God, and from whose eyes flow tears of love—he has indeed revealed his last birth" (ibid., p.94).
8. Swami Nikhilananda, op.cit., p.912.
9. See Romain Rolland, *The Life of Ramakrishna* (Almora: Advaita Ashram, 1931) p.263, fn.1.
10. Swami Nikhilananda, op.cit., p.153, emphasis added.
11. What Ramakrishna seems to have believed in on the basis of his personal experience is not *reincarnation* but *incarnation*. He declared himself to be an incarnation in the last days of life, when he was suffering from cancer of the throat. At one point during this illness Narendra, who later on became known as Swami Vivekananda said to himself, 'If in the midst of this racking physical pain he declares his Godhead, then only shall I accept him as an incarnation of God'. He was alone by the bedside of the Master. It was a passing thought, but the Master smiled. Gathering his remaining strength he distinctly said, 'He who was Rama and Krishna is now, in this body, Ramakrishna-but not in your Vedantic sense'" (Swami Nikhilananda, op.cit., p.72). Ramakrishna also regarded Narendra himself as an incarnation of *Nara*, as is clear from Swami Vivekananda's own account of one of their earlier meetings: "The next moment he stood before me with folded palms and, showing me the regard due to a god, went on saying, 'I know, my lord, you are the ancient Rishi *Nara*, a part of *Narayana*, who has incarnated himself this time, to remove the miseries and sufferings of humanity'" (See Swami Saradananda, *Sri Ramakrishna: The Great Master* [Madras: Sri Ramakrishna Math, 1952] p.177). Also see Nalini Devdas, *Sri Ramakrishna* (Bangalore: The Christian Institute for the Study of Religion and Society, 1966) pp.92-104. When Ramakrishna *talks of his being born again he speaks of it as an incarnation*. Thus one evening he said, "The illness of my body is caused by the sins of those who come and touch my feet. I purify the sinners by taking their sins upon myself and suffering for them. He who was Rama, who was Krishna, Buddha, Christ and Chaitanya has now become Ramakrishna. Blessed are those who know this truth. My divine Mother has shown me that the photograph of this body will be kept upon altars and be worshipped in different houses as the pictures of other Avataras are worshipped. My divine Mother has also shown me that I shall have to come back again and that my next incarnation will be in the West" *(Memoirs of Ramakrishna* [Calcutta: Ramakrishna Vedanta Math, 1957] pp.339 340).

13

SOME UNRESOLVED QUESTIONS REGARDING THE RETURN OF RAMAKRISHNA

I

Ramakrishna will be reborn. Swami Vivekananda, his pre-eminent disciple, tells us that "He said he would come back once more with us. Then, I think he will embrace *Videha-mukti* (Absolute emancipation)".[1] Not only do we have this on the authority of his best-known disciple, he is said to have declared himself:

> My divine Mother has shown me that the photograph of this body will be kept upon altars and be worshipped in different houses as the pictures of other *Avataras* are worshipped. *My divine Mother has also shown me that I shall have to come back again and that my next incarnation shall be in the West.*[2]

If it, then, be accepted that Ramakrishna will be reborn, as seems to be clearly indicated by the statements of both the Master and his disciple, then two questions arise: (1) where will the rebirth take place? and (2) how is this rebirth to be reconciled with the spiritual attainments of Ramakrishna for typically the sage is freed from *samsara*.

II

In view of the words of Ramakrishna *quoted above*, the answer to the first question does not seem to be in doubt. Indeed the question itself appears redundant, as it appears to have been already answered.

It turns out, however, that this is only one account of the locale of his next birth, namely, the West. As the word is cited in capital letters, one would take it to mean what we usually mean by the West. On this view then, Ramakrishna is destined to be born in North America or Western Europe, even though one might feel that this is carrying "Westernization" a bit too far.

Another account, however, depicts Ramakrishna as making the following statement:

Those belonging to the inner circle will not attain liberation. I shall have to assume a human body again, in a northwesterly direction.[3]

The fact of Ramakrishna's rebirth is reasserted in this passage but the direction involved has changed from West to north-west.

This fact that Ramakrishna predicted his own rebirth is noticed by Romain Rolland who commented:

As a curious fact I note here that Ramakrishna said, pointing to the north-west, that after two hundred years he would be reincarnated there.[4]

And then Romain Rolland parenthetically asks: Russia?[5]

So Ramakrishna will be reborn, after about two hundred years, either in the USA(?) or in the USSR(?).[6]

III

Now the next question. How come a sage like Ramakrishna will be reborn?

Before this question is answered, it should be borne in mind that Ramakrishna did not seem to have personally believed in rebirth-personally, that is, on the basis of personal experience.[7] This point has been discussed in Chapter [12].

One needs here, perhaps, to distinguish between two notions of reincarnation (1) one of the ordinary soul reborn in the process of *samsara* and (2) of God or divine beings who Incarnate themselves. The first may be referred to as reincarnation and the latter as re-Incarnation. In contrast to his hesitation about reincarnation, Ramakrishna is quite forthright when it comes to re-Incarnation. Romain Rolland tells us:

> Two days before his death, in answer to Naren's unspoken desire to drag from him the avowal he was so loth to make, Ramakrishna said, "He who was Rama, and who was Krishna is now Ramakrishna in this body lying here!"

> But. he added, "Not in your Vedantic sense". (That is to say, not merely in the sense of identity with the Absolute, but in the sense of Incarnation).[8]

It is clear, then, that for Ramakrishna, his next birth, wherever it may be, will not be a mere rebirth but a re-Incarnation.

The postcript-like comment by Ramakrishna, however, that he is not an Incarnation in the Vedantic (= *advaitic)* sense needs to be examined. It creates room for suggesting that from the point of view of *bhakti* Ramakrishna's "second coming" may be an Incarnation but it may not be so from the point of view of *Jnana*. For *Advaita Vedanta*, although it provides room for "coming down" from the non-dual experience in the course of one life in its distinction between *Jivanmukti* and *Videhamukti* it does not normally leave room for a rebirth.[9] To be reborn is *not* to have been fully liberated. Ramakrishna pointed out that:

> The forms and aspects of God disappear when one discriminates in accordance with *Vedanta* philosophy. The ultimate conclusion of such discrimination is that *Brahman* alone is real and this world of names and forms is illusory. It is possible for a man to see the forms of God, or to think of Him as a person, only so long as he is conscious that he is a devotee. *From the standpoint of discrimination this 'ego of a devotee' keeps him a little away from God.*[10]

In the *advaitic* scheme, it seems all beings undergo rebirth. By declaring that he is not an Incarnation in the *Vedantic (=advaitic)* sense, does Ramakrishna suggest that he is to be treated as one in the *bhaktic* sense? Then the question arises: is he re-Incarnating as God or as a devotee, for "according to Ramakrishna, those *bhaktas* who keep the trace of distinction-the feeling of 'I' and 'Thou', *bhakta* and *Bhagvan*—deliberately renounce liberation and stay within the wheel of *samsara* to enjoy the love of God. But through love they can reach non-difference and transcend the cycle of rebirths".[11] At times Ramakrishna speaks like a *bhakta.*[12] On other occasions, however, he seems to speak like an *Avatara.*[13]

To conclude: Ramakrishna will be reborn, though we do not quite know where; and his rebirth will be in the nature of re-Incarnation, though we do not quite know whether this descent is of a *bhakta* or of *Bhagvan*.

References

1. *The Complete Works of Swami Vivekananda* Vol.VII (Calcutta: Advaita Ashrama, 1969) p.413, emphasis added.
2. Swami Abhedananda, Memoirs*of Ramakrishna* (Calcutta: Ramakrishna Vedanta Math, 1957) pp.339-340, emphasis added.
3. Swami Nikhilananda, *The Gospel of Sri Ramakrishna* (New York: Ramakrishna - Vivekananda Center, 1942) p.829.
4. Romain Rolland, *The Life of Ramakrishna* (Calcutta: Advaita Ashrama, 1960) p.174 fn.2.
5. ibid. Does Nikhilananda's account designate the place as the West because *he* was carrying out his missionary activity in the West?
6. These conflicting accounts raise some doubts about the claims of authentic reporting made by some writers about Ramakrishna (see Christopher Isherwood, *Ramakrishna and His Disciples* [London: Methuen & Co. Ltd., 1965] p.2) and remind one of the Gospel divergences.
7. Swami Nikhilananda, op. cit., p.153, emphasis added; see also p.912.
8. Romain Rolland, op. cit., p.273, fn.2.
9. Eliot Deutsch and J.A.B. van Buitenen, *A Source Book of Advaita Vedanta* (Honolulu: University of Hawaii Press, 1971) p.312.
10. Swami Nikhilananda, op. cit., p.150, emphasis added.
11. Nalini Devdas, *Sri Ramakrishna* (Bangalore: The Christian Institute for the Study of Religion and Society, 1966) p.18.
12. Swami Nikhilananda, op. cit., p.481.
13. Romain Rolland, op. cit., p.273, fn.2.

14

WAS RAMAKRISHNA AN AUTHENTIC GURU?

I

Critics of the Ramakrishna movement have argued that the movement or mission is not authentic because it is not a lineal descendant of any of the established monastic orders. It is alleged that the Ramakrishna Mission has tried to artificially graft itself on the *Dasanami* order by virtue of the fact that Totapuri, who initiated Ramakrishna into *Advaita sadhana* belonged to that Order.

The purpose of this essay is to examine the contention that Ramakrishna on that account cannot be regarded as an authentic guru.

II

In order to judge whether Ramakrishna was an authentic Guru or not, one must identify the criteria within the tradition for determining such authenticity. These criteria are provided by the *Upanisads* which state that a guru should be (1) a *srotriya* or well-versed in the Vedas and (2) *brahmanistha,* established in the experience of *Brahman.*[1]

Ramakrishna's experience of *Nirguna Brahman* is well-known[2] and accounts of Ramakrishna's experiences of *Saguna Brahman* are varied. The following one concerns his experience of his favourite goddess, Kali:

> The Master's worship, meditation, etc., underwent novel changes. It is difficult to explain to others that wonderful state of complete absorption in Her.... Whenever one saw him, one thought he had

> merged his little will and the little ego in Her will, Who was the source of all wills, and did everything as if he was completely an instrument in Her hand, praying in his heart of hearts, 'Mother, my only refuge, kindly make me, Thy boy, say and do what I should.' ... Although in the world, he was not of it.... The universal Mother's form, consisting of pure consciousness and bliss, was now known to him as the only reality....
>
> Now he saw, even at times other than those of worship and meditation, the full figure of the effulgent Mother, smiling and speaking, guiding him, accompanying him and saying, 'Do this, don't do that'.
>
> ... When he offered cooked food etc., to the Mother,... [She] sat down to take the offerings in Her very person, illumining the temple with the effulgence of Her holy person....[33]

One might ask at this point: the evidence presented goes to show that Ramakrishna had experienced *Brahman*. But what is there to prove that he was firmly established in that experience? The biographical details of his life leave little room for scepticism on this point.[4]

III

It is thus established that Ramakrishna was a *brahmanistha*. Was he a *srotriya*?

Before answering the question, it would be useful to examine the significance of the requirement. Why should a guru be a *srotriya*? A little reflection will suggest that this requirement has to do with the ability of the Guru to communicate with his disciples. Since in Vedic times it was the Vedas which represented the shared intellectual tradition of the people, the Guru had to be a *srotriya* or well-versed in the Vedas. But if it be granted that the real purpose of this requirement is communicability then Ramakrishna was eminently successful in this. Technically he was not a *srotriya* in the sense of being a traditional Vedic scholar but functionally he was, in the sense of being able to communicate his spiritual insights and experiences to his disciples. He was illiterate, but he was capable of communicating his ideas, ideals and experiences not merely to the illiterate but also to the literate circle of followers that gathered round him.

IV

One more point remains to be covered: he did not belong to any *guru-parampara* and therefore cannot be regarded as an authentic guru. The argument again is technically forceful but seems weak if examined critically. Most of the *guru-paramparas* trace themselves to an *adi- guru.*[5] What if in Ramakrishna we have such an *adi-guru.*? The Hare Krishna movement traces itself back to Chaitanya and also beyond, but sets more store by its lineal descent from him. The *Advaita* monastic orders trace themselves back to Shankara, the *Visistadvaita* trace themselves back to Ramanuja. What if we have in the Ramakrishna mission the emergence of a new *sampradaya*, notwithstanding its claim to universality?

V

To conclude: Ramakrishna has to be accepted as an authentic guru by the criteria developed within the tradition once they are critically rather than mechanically applied to his case.

References

1. *Mundakopanisad* II.2.12.
2. See Claude Alan Stark, *God of All* (Cape Cod Mass: Claude Stark Inc., 1974) p.61, wherein the passage is cited and clearly identified as relating to *nirvikalpa samadhi*.
3. ibid., p.29.
4. See Swami Jagadananda, tr., *Sri Ramakrishna The Great Master by Swami Saradananda* (Madras: Sri Ramakrishna Math, 1963) passim; Swami Nikhilananda, tr., *The Gospel of Sri Ramakrishna* (New York: Ramakrishna-Vivekananda Center, 1942).
5. Eliot Deutsch & J.A.B. van Burtenen, *A Source Book of Advaita Vedanta* (Honolulu: The University Press of Hawaii, 1971) p.65.

MAX MUELLER'S *RAMAKRISHNA: HIS LIFE AND SAYINGS:* A STUDY IN MOTIVATION

I

Why did Max Mueller write the first major biography of Ramakrishna in English?[1]

II

An examination of the Preface Max Mueller wrote to the book suggests a complex network of motivation. One may begin to unravel this complexity by identifying individual motivational strands as far as possible. Max Mueller wrote his book:

(1) Because "the name Ramakrishna has lately so often been mentioned in Indian, American and English newspapers".[2]

(2) Because of this he has become important enough for "many who take an interest in the intellectual and moral state of India".[3]

(3) Because there are a few "to whom the growth of philosophy and religion, whether at home and abroad, can never be a matter of indifference".[4]

(4) Because the utterances of a Hindu *Mahatma* "which sway large multitudes which gather round them in their own country, cannot fail to engage our attention and sympathy".[5]

(5) Because his doctrines "are being spread by zealous advocates not only in India, but in America also, nay even in England".[6]

(6) Because "a better knowledge of the teachings of one of 'the Sanyasins' seems certainly desirable... for the statesmen who have to deal with various classes of Indian society."[7]

(7) Because 'a better knowledge' of the teachings of one of the Sanyasins "seems certainly desirable ... for the missionaries who are anxious to understand and to influence the inhabitants of that country".[8]

(8) Because "students of philosophy and religion ... ought to know how the most ancient philosophy of the world, the *Vedanta*, is taught at present day by the *Bhaktas*, that is 'the friends and devoted lovers of God' and continues to exercise its powerful influence, not only on a few philosophers, but upon the large masses of what has always been called a country of philosophers."[9]

(9) Because a "country permeated by such thoughts as uttered by Ramakrishna cannot possibly be looked upon as a country of ignorant idolators to be converted by the same methods which are applicable to the races of Central Africa".[10]

III

These various motives may be identified as connected with:

(a) Ramakrishna, his international fame (no.1), his influence in India (no.4), and the missionary activity of his disciples (no.5);

(b) an intellectual interest in India (no.2);

(c) broad philosophical reasons (nos 3 and 8);

(d) administrative reasons (no.6) and

(e) missionary reasons (nos 7 and 9).

IV

Thus there were several motives which led Max Mueller to write his book. Among these the following seem to have been prominent:

(a) the national influence (no.4), the international fame (no.1) of Ramakrishna, and the missionary activity conducted by his followers (no.5);

(b) the broad philosophical reasons of growth of religion and philosophy (no.3) and specially the way the philosophical tradition functions in India (nos 8 and 2);

(c) narrow political reasons (no.6); and

(d) the fact that if one wanted to be an effective missionary in India (no.7), one should realize that a strategy different than the one used in Central Africa will have to be followed (no.9).

The various elements are perhaps too closely connected to be disentangled but it is clear that the motivation represents a curious blend of internationalism and imperialism; of genuine religiosity and missionary Christianity.

References

1. See Romain Rolland, *The Life of Ramakrishna* (Calcutta: Advaita Ashrama, 1960) pp.297-302; Christopher Isherwood, *Ramakrishna and His Disciples* (London: Methuen & Co. Ltd., 1965) pp.335-337, etc.
2. F. Max Mueller, *Ramakrishna: His Life and Sayings* (London: Longmans, Green & Co., 1898) p.v.
3. ibid.
4. ibid.
5. ibid., p.vi.
6. ibid.
7. ibid., p.vii.
8. ibid.
9. ibid.
10. ibid.

16

MAX MUELLER AS A BIOGRAPHER OF RAMAKRISHNA

Of the several biographical accounts of the life of Ramakrishna, the first major work to appear in English was Max Mueller's *Ramakrishna: His Life and Sayings*[1] On further investigation it turns out, however, that Max Mueller can be called the author of the parts directly dealing with Ramakrishna's life only in an indirect sense. Thus before commencing an account of the life of Ramakrishna, Max Mueller tells us:

> We possess indeed full accounts of his life, but they are often so strangely exaggerated, nay so contradictory, that it seemed almost hopeless to form a correct and true idea of his earthly career and his character.[1] I applied therefore to one of his most eminent pupils, Vivekananda, asking him to write down for me what he could tell from his own knowledge of his venerable teacher, and I received from him a full description of his Master's life. It will be easily seen, however, that even this account is not quite free from traditional elements. If I give as much as possible unaltered, I have reason for it.[2]

It is clear from this passage that (1) the account given by Max Mueller is really the account of Ramakrishna's life as given by Swami Vivekananda, and (2) that it is not an "unaltered" account, as Max Mueller implies in the last sentence in the passage cited above.[3]

Thus Max Mueller is not really the *author* so much as the *redactor* of the life of Ramakrishna. The purpose of this essay is to examine, not his motivation for presenting the life of Ramakrishna but the pattern of the presentation.

II

Although normally a paper is supposed to build up to a conclusion which is presented at the end, in this case it seems a more useful procedure to present the conclusion first and then adduce the evidentiary material in support of it.

It seems that Max Mueller followed what may be called an *apologetic*[4] *pattern* in presenting the life of Ramakrishna. This becomes clear from an examination of his prefatory remarks as well as from incidental remarks scattered throughout the book.

In the Preface, for instance , Max Mueller refers to the accounts of *Sannyasins* and *Hathayogins* emanating from India and comments:

> Though some of the stories of these martyrs of the flesh and of the spirit may be exaggerated, enough remains of real facts to rouse at all events our curiosity. When some of the true sanyasins, however, devote their thoughts and meditations to philosophical and religious problems, their utterances, which sway large multitudes that gather round them in their own country, cannot fail to engage our attention and sympathy, particularly if, as in the case of Ramakrishna, their doctrines are being spread by zealous advocates not only in India, but in America also, nay in England itself.[5]

Max Mueller here seems to suggest that Ramakrisna may be placed in the latter category and therefore should be taken seriously[6] This remark may also be seen in the broader background of his position, that from these genuine saints presumably, as from Indian lore in general, "there was something real, something worth knowing even for us, the students of Plato and Aristotle, Kant and Hegel, in Europe."[7]

This apologetic attitude of Max Mueller, this tendency to defend Ramakrishna in front of a European audience, not merely surfaces in but sometimes even influences the narration of the life of Ramakrishna.[8] Thus, while speaking of Ramakrishna's experience of *madhura bhava* he writes:

> The highest point of love is reached when the human soul can love his God as a wife loves her husband. The sphepherders of *Braja* had this sort of love towards the divine *Krishna,* and there was no thought of any carnal relationship.[9]

The fact is that the women of Vraja left their husbands and went after Krishna,[10] which is, of course, shocking from the point of view of normal ethics. This seems to have led him, or (Swami Vivekananda?) to omit the suggestion of illicit amour involved here.

After providing Vivekananda's version of Ramakrishna's life, Max Mueller also takes into account some criticisms of Ramakrishna made by Mozoomdar, two of them being that (1) Ramakrishna used filthy language and (2) maltreated his wife. On both these counts Max Mueller acquits Ramakrisna. On the issue of filthy language he writes:

> In a country where certain classes of men are allowed to walk about in public stark naked, language is not too likely to veil what with us requires to be veiled. There is, however, a great difference between what is filthy and what is meant to be filthy. I doubt whether the charge of intentional filthiness or obscenity, which has been brought against writers like Zola, could be brought, or has been brought against Ramakrishna.[11]

As for the maltreatment of his wife, it consisted essentially of the fact that the marriage was not consummated. Max Mueller remarks:

> It is strange that a man of Mozoomdar's knowledge and experience should have considered the resolve of Ramakrishna's wife to live with him as a *sanyasini* as barbarous treatment. She herself evidently did not think so, nor have I heard of any other cruelties on the part of her husband. If she was satisfied with her life, who has any right to complain; is love between husband and wife really impossible without the procreation of children? We must learn to believe in Hindu honesty, however incredulous we might justly be on such matters in our own country.[12]

III

It is clear, therefore, that Max Mueller's account of Ramakrishna was really *meant for a Western audience* and that in keeping with the nature of this audience Mueller either did not present, or if he did present, then took steps to vindicate such material as might do what he would have considered unjust harm to a proper acceptance of the life of Ramakrishna[13] on its own merits

References

1. See Romain Rolland, *The Life of Ramakrishna* (Calcutta: Advaita Ashrama, 1960) pp. 297-302; Christopher Isherwood, *Ramakrishna and His Disciples* (London: Methuen & Co. Ltd., 1965) pp.335-337, etc.
2. Max Mueller, *Ramakrishna: His Life and Sayings* (London: Longmans, Green & Co., 1898) p.24.
3. "Vivekananda himself is a man who knows England and America well, and perfectly understood what I meant. Yet even his unvarnished description of his Master discloses here and there the clear traces of what I call the Dialogic Process, and the irrepressible miraculising tendencies of devoted disciples... I have also, so far as I could, consulted another account of the life of Ramakrishna published in the late numbers of the *Brahmavadin*. But I am sorry to say that this account stops at No 19, and has not been continued" (ibid., p.30)., Diacritics omitted, emphasis added.
4. The use of the word 'apologist' in one sense may be misleading. Apologists are often accused of distorting the truth to present their heroes in a better light. In Max Mueller's case, however, given the legendary nature of the accounts associated with Indian sages, his apologetics consisted in presenting a historically trustworthy account. Thus at the conclusion of the account of Ramakrishna's life he remarks: " This is all that Vivekananda sent me when I had asked him to write down whatever he could gather from his own memory and from communication with Ramakrishna's other disciples. I had warned him repeatedly not to send me mere fables, such as I had read about his guru in several Indian periodicals and I believe he fully understood what I meant " (ibid., p.60).
5. ibid., p.vi. Diacritics omitted
6. For instance, consider the title of his article 'A Real Mahatman' which "appeared in the August number, 1986, of the *Nineteenth Century*" (ibid., p.l.).
7. ibid., p.2.
8. Unless this is a case where Swami Vivekananda has been presented unaltered.
9. Max Mueller, op.cit., p.50. Emphasis added. The love of a wife for her husband and a woman for her paramour is sometimes distinguished in *Vaishnava* theology (see T.M.P Madhadevan, *Outlines of Hinduism*, (Bombay: Chetana Ltd., (1960) p.91. Here this distinction has been obscured. Has this anything to do with Max Mueller's advice to Ramakrishna's followers that "nothing would be more likely to lower their master and their own work in the eyes of serious people than the slightest moral laxity on their part or a defence of any such laxity on the ground that a Gnanin, a Knower, is above morality"? (ibid., p.92. Diacritics omitted).
10. See Thomas J. Hopkins, *The Hindu Religious Tradition* (Belmont, California: Dickenson Publishing Co. Ltd, 1971) p.105.
11. Max Mueller, op.cit., pp.62-63
12. ibid., p.65. In the same vein, Max Mueller adds elsewhere, "If, as we are told, he did not show sufficient moral abhorrence of prostitutes, he does not stand quite alone in this among the founders of religion. If he did not 'honour the principle of teetotalism according to Western notions', no one, so far as I know, has accused him of any excess in drinking" (ibid., p.67).
13. It should not be inferred from this that Max Mueller, in any sense, had no reservations about Ramakrishna. He remarks on cases of apparent exercise of spiritual power on others by Ramakrishna parenthetically as "evidently case of

hypnosis" (ibid., p.38). In this case he comes close to J.H. Farquhar who comments: "It is also said that he (Ramakrishna) had the power to induce *samadhi* in others. This trance is clearly a form of hypnotism" *(Modern Religious Movements in India* [New York: The Macmillan Co., 1915] p.189).

17

THE LOGIA OF RAMAKRISHNA AND THE APOLOGIA OF MAX MUELLER

Ramakrishna: His Life and Sayings by Max Mueller, really consists of two parts as the title itself suggests: his life and his sayings.

Max Mueller tells us that he "received a complete collection" of Ramakrishna's sayings from Ramakrishna's "own pupil, Vivekananda, well-known by his missionary labours in the United States and England. I give them as they were sent to me, with such corrections only as seemed absolutely necessary."[1]

The purpose of this essay is to draw attention to the apologia of Max Mueller which accompanies the logia of Ramakrishna, which indicates that Max Mueller's motives in presenting these sayings were not merely historical but also apologetic.[2]

Max Mueller points out how he overcame his reluctance to present the sayings of Ramakrishna *as they are* in the interest of historical truth. Though Max Mueller presumably presents the picture as it is, it is presented in an apologetic framework. The overall decision to present the sayings is influenced by the consideration that they help to show certain aspects of *Vedanta*. Thus Max Mueller writes:

> To my mind these sayings, the good, the bad, and the indifferent, are interesting because they represent an important phase of thought, an attempt to give prominence to the devotional and practical side of *Vedanta*, and because they show the compatibility of *Vedanta* with other religions. They will make it clear that the *Vedanta* also possesses a morality of its own, which may seem too high and too spiritual for ordinary mortals, but which in India has done good, is doing good, and may continue to do good for centuries to come.[3]

The remarks Max Mueller makes around some of the individual sayings are quite revealing in this respect. Many of these remarks are made in the Preface, wherein he shows the awareness that some of the sayings "may sound strange to our ears, nay even offensive."[4] He picks out some of them to illustrate his point:

No.89: "Why does the God-lover find such pleasure in addressing the Deity as Mother? Because the child is more free with its mother, and consequently she is dearer to the child than any one else".

Max Mueller points out that "the conception of the Deity as the Divine Mother is apt to startle us" but hopes that "we can understand" what Ramakrishna really meant.[5]

No.104: "A true devotee who has drunk deep of Divine Love is like a veritable drunkard, and, as such, cannot always observe the rules of propriety".

No.92: "What is the strength of a devotee? He is a child of God, and tears are his greatest strength".

Max Mueller regards these statements as made "in excuse" of the fact that "sometimes the language which these Hindu devotees use of the Deity must appear to us too familiar, nay even irreverent."[6]

No.172: "The Knowledge of God may be likened to a man, while the love of God is like a woman. Knowledge has entry only up to the outer rooms of God, but no one can enter into the inner mysteries of God save a lover, for a woman has access even into the harem of the Almighty".

Max Mueller takes pains to point out that the "saying will certainly jar in our ears", "unless we remember that harem means originally no more than a sacred and guarded place".

Max Mueller also presents certain utterances which show "the exalted nature of his faith" such as:

No.201: "He who has faith has all, and he who wants faith wants all".

No.288: "As a lamp does not burn without oil, so a man cannot live without God".

No.215 "God is in all men, but all men are not in God: that is the reason why they suffer."[7]

Soon Max Mueller draws the conclusion from these and similar sayings that:

> The consciousness of the Divine in man is there, and is shared by all, *even by those who seem to worship idols.*[8]

From these sayings of Ramakrishna, Max Mueller also seems to take the cue for the harmony of religions of which Ramakrishna himself was so acutely aware.

> This constant sense of the presence of God is indeed the common ground on which we may hope that in time not too distant, the great temple of future will be erected, in which Hindus and non - Hindus may join hands and hearts in worshipping the same Supreme Spirit who is not far from every one of us, for in Him we live and move and have our being.[9]

Max Mueller's defence of Ramakrishna here reaches such a pitch that one wonders for a moment whether the last statement is his or Ramakrishna's!

References

1. Max Mueller, *Ramakrishna: His Life and Sayings* (London: Longmans, Green & Co., 1898) p.96. He adds later on that "Some more of Ramakrishna's sayings have been sent to me lately, but their publication will have to wait for another opportunity" (ibid., p.98, fn.1).
2. For the need for such studies see Alexander Lipski, "Some Methodological Problems Encountered by Western Scholars in the Study of Modern Indian Religious History", *Journal of Indian History*, Vol.48 (1970) p.275ff
3. Max Mueller, op.cit., p.97.
4. ibid., p.vii.
5. ibid., pp .vii-viii.
6. ibid.,p.viii.
7. ibid., p.ix.
8. ibid., emphasis added.
9. ibid., p.x.

18

MAX MUELLER AND ROMAIN ROLLAND AS BIOGRAPHERS OF SRI RAMAKRISHNA

I

Several biographies of Ramakrishna Paramahamsa are now available.[1] Out of these several biographies some have been written by Western writers and among these, those by Max Mueller[2] and Romain Rolland[3] are fairly well known. The purpose of this essay is to draw attention to a curious contrast in the approaches of these scholars towards the application of modern science to the type of experiences associated with Sri Ramakrishna.

II

In the Preface to his book on Ramakrishna, the first major biography by a Western scholar, Max Mueller remarks:

> We need not fear that the Sanyasins of India will ever find followers or imitators in Europe, nor would it be all desirable that they should, not even for the sake of Psychic Research, or for experiments in Physico-psychological Laboratories.[4]

These remarks are a bit curious especially in view of the fact that Max Mueller does attempt to offer scientific or natural explanations for some of the experiences associated with Ramakrishna. More specifically, he certainly does so on two occasions parenthetically. He does it the first time when describing a mystical experience of Ramakrishna in early childhood:

> Before he reached his teens, he was walking in the fields one day. The sky was very clear and blue, and he saw a flight of white cranes moving along it. The contrast of colours was so very beautiful and dazzling to his imagination and produced such thoughts in him, that he fell down in a trance. (This would admit of a very natural pathological explanation, and may therefore be perfectly true, though it would easily lend itself to further poetical expansion).[5]

The other case is more intriguing. In this case Ramakrishna induced a trance in a lady "of a noble family, the wife of a gentleman and mother of five or six children", "a perfect stranger" to establish that "some women were born with all the qualities of a *devi.*"[6] But the trance did not:

> ....leave her for some hours, and he got frightened at the thought that her husband would accuse him of some black magic. He began therefore, to pray to his mother *Kali* [the goddess] to bring her back to her senses. By and by she came to herself, and when she opened her eyes they were quite red, and she looked as if she were quite drunk. Her attendants had to support her while she got into a carriage, then she drove back home. This is one of many instances of the same kind (evidently cases of hypnosis).[7]

In both these instances, Max Mueller suggests rational or "scientific" explanations himself, which is somewhat odd in view of his disinclination, expressed earlier, towards subjecting such incidents to scientific scrutiny. Later he writes thus about the passing away of Ramakrishna:

> He was undaunted and remained cheerful as ever, till on August 16, 1886, at 10 o'clock in the night, he entered into *Samadhi* from which he never returned. His disciples took it at first to *be an ordinary Samadhi, such as he used to have every day, during which the best of doctors could not find any pulsation or beating of the heart*; but, alas, they were mistaken.[8]

It is clear that Max Mueller was aware of the apparent disappearance of cardiac activity in a state of *Samadhi*, but this does not seem to have piqued his scientific curiosity.

III

The attitude of Romain Rolland, on the other hand, is quite different. Referring to the illness of Ramakrishna he points out that Dr Sarkar, who attended on him was "a rationalist who did not share the religious views of Ramakrishna"[9]and that:

> He was present during several ecstasies and studied them from a medical point of view. A study of his notes would be of great interest to European science. It is known that a stethoscopic examination of the heart and the condition of the eyes during *samadhi* show all the symptoms of the condition of death.[10]

Max Mueller refers to the effectiveness of Ramakrishna's "touch" along with his other psychic powers".[11] Romain Rolland does the same but then adds:

> For *scientific* men, who study psycho-physiological problems, it is noteworthy that these "touches", which provoked in the subjects concerned, immediate experience of changed conditions, were nearly always (if not always) produced when Ramakrishna was in a state of semi-consciousness or of complete hypnosis. There was therefore nothing in them analogous to calculated action of the will independent of the energies governed by it. It might almost be described as a forced descent of another into the abyss he had first descended himself.[12]

IV

It is clear, therefore, that while Romain Rolland seems to encourage the scientific investigation of the phenomena associated with Ramakrishna, Max Mueller does not do so. At first sight it might appear that he is held back by the fact that he is only narrating what Swami Vivekananda had reported to him about his Master. But it is quite obvious that Max Mueller's attitude is not one of naive acceptance of Vivekananda's account.[13] In fact, it might be said that both Max Mueller and Romain Rolland are critical,[14] if sympathetic biographers, but that in the case of Max Mueller this criticality is largely historiographic while in the case of Romain Rolland it takes on an added dimension of being scientific.[15]

References

1. See Christopher Isherwood, *Ramakrishna and His Disciples* (London: Methuen & Co. Ltd., 1965) etc.
2. Max Mueller, *Ramakrishna: His Life and Sayings* (London: Longmans, Green & Co., 1898).
3. Romain Rolland, *The Life of Ramakrishna* (Calcutta: Advaita Ashrama, 1960)
4. Max Mueller,op.cit.,pp.vi-vii.Diacritics omitted.
5. Max Mueller, op.cit., p.34.
6. ibid., p.37. Diacritics omitted.
7. ibid., p.38. Diacritics omitted.
8. Max Mueller, op.cit., p.57; emphasis added.
9. Romain Rolland, op.cit., p.262.
10. Romain Rolland, op.cit., p.263.
11. Max Mueller, op.cit., p.54.
12. Romain Rolland, op.cit., p.240, fn.1, emphasis added.
13. Max Mueller, op. cit., pp. 24-30,59-61.
14. See Romain Rolland, p.197, fn.1; p.225, fn.2; p.149, fn.2, etc.
15. ibid.See footnotes on pp.150, 151, 155, 216, 240, 263. In this respect Romain Rolland may have been overgenerous in his annotation on Max Mueller's work in the bibliography to his own work. He comments: "Max Mueller knew Vivekananda personally in England; and he asked him to give him a complete account of the life of his Master. His small work is therefore based on first hand evidence, and he uses it with his broad and clear critical spirit , in which are allied the *scientific* exigencies of the West and a generous understanding of all forms of thought" (ibid., p.301, emphasis added). The sense in which the word scientific is used here is not clear; but the sense in which it may *not* be applied to Max Mueller's account is, I think, clear from the essay.

19

FOLKLORE AND MODERNITY: THE CASE OF RAMAKRISHNA

I

The term Folklore was coined in 1846 by W.J. Thomas in order to replace the term "popular antiquities".[1] The term is now generally used to denote "Traditional Beliefs, Customs, Stories, Songs and Sayings current among backward peoples, or retained by the uncultured masses of more advanced peoples."[2] For the purposes of this paper it is suggested that the definition be modified to read "....or retained by the *illiterate* of more advanced peoples". With this modification, one may proceed to discuss the effect of familiarity with modern mechanical inventions on Folklore.

Folklore, as defined above, can then be taken to include the Sayings of an illiterate mystic such as Ramakrishna Paramahamsa. A collection of such Sayings forms part of the first biography of the saint in English by F. Max Mueller,[3] who wrote his work mainly on the basis of the material supplied to him by that well-known disciple of Ramakrishna, Swami Vivekananda.[4] An analysis of these Sayings contained in Max Mueller's work shows an interesting stage in the influence of modernity—by which here is meant modern mechanical inventions and ideas—on Folklore.

Max Mueller remarks that Ramakrishna's "sayings or Logia were collected and written down by his pupils in Bengali; some were translated into Sanskrit and into English. There are many that remind us of old Sanskrit sayings, of which there are several collections, all, however, in metrical form. The sayings of Ramakrishna are different because they are in prose, uttered evidently on the spur of the moment, and tinged here and there with European ideas which must have reached Ramakrishna through his intercourse with Anglo-Indians and not from books, as he was ignorant of English."[5] Elsewhere too, speaking of Ramakrishna and

Keshub Chunder Sen, his contemporary, Max Mueller remarks : "What is curious, however, both in Keshub Chunder Sen, and in Ramakrishna's utterances, is the admixture of European ideas. Neither the one nor the other would have spoken as they did, before the English Government began its educational work in India. The bulk of their teaching is, no doubt Indian to the backbone. It is the old Indian philosophy, properly called *Vedanta* or the highest goal of the Veda, but there is clearly a sprinkling, and sometimes far more than a mere sprinkling, of European thoughts in Keshub's writings; and we often meet with quite unexpected references to European subjects, not excluding railways and gas, in the sayings of Ramakrishna."[6]

Even before one turns to the Sayings, however, the intrusion of the modern West even into the mystic's life is attested to. We read for instance that:

> 'Sometimes', he said, I would sit by the Ganges, with some gold and silver coins and a heap of rubbish by my side, and taking some coins in my right hand and a handful of rubbish in the left, I would tell my soul, "My soul! this is what the world calls *money, impressed with the queen's face*. It has the power of bringing you rice and vegetables, of feeding the poor, of building houses, and doing all that the world calls great, but it can never help them to realise the ever-existent knowledge and bliss, the *Brahman*. Regard it, therefore, as rubbish." Then mixing the coins and the rubbish in my hands, while repeating all the time, :"money is rubbish, money is rubbish", I lost all perception of difference between the two in my mind, and threw them both into the Ganges. No wonder people took me for mad.[7]

The coins obviously bore the impression of Queen Victoria and Ramakrishna apparently knew it to be so. It is, however, in his Sayings that the interaction of modern inventions and ideas with folk material is most obvious.[8]

One of the metaphors Ramakrishna uses is obviously drawn from the sight and perhaps experience of the medicinal injection. Thus he comments:

> How doth the Lord dwell in the body? He dwells in the body like the plug of a syringe, i.e.in the body, and yet apart from it.[9]

And again:

> How does the soul stay in the body? As the piston stays in a syringe.

The following two sayings seem to allude to both the political and religious influences operating in his times—the Viceroy as the representative of the Royal House in India and Christ as representative of Protestant Christianity.

> The *Avatara* or Saviour is the messenger of God. He is like the Viceroy of a mighty monarch. As when there is some disturbance in a far-off province, the king sends his viceroy to quell it; so whenever there is any waning of religion in any part of the world, God sends His *Avatara* there.
> It is one and the same *Avatara* that, having plunged into the ocean of life, rises up in one place and is known as Krishna, and diving again rises in another place and is known as Christ.[11]

Among the mechanical changes of the landscape wrought by the British the Railways figure prominently. Thus the saint remarks:

> A locomotive engine reaches the destination itself, and also draws and takes with it a long train of loaded wagons. So likewise act the Saviours. They carry multitudes of men, heavily laden with the cares and sorrows of the world, to the feet of the Almighty.[12]

And again:

> The locomotive engine easily drags along a train of heavily-laden carriages. So the loving children of God, firm in their faith and devotion in Him, feel no trouble in passing through all the worries and anxieties of life, and leading many men along with them to God.[13]

There are references to the steamer as well:

> As a large and powerful steamer moves swiftly over the waters, towing rafts and barges in its wake, so when a Saviour descends, He easily carries thousands across the ocean of *Maya* (illusion).[14]

There is a reference to gas in the Sayings. Thus Ramakrishna remarks:

> The light of the gas illumines various localities with various intensities. But the life of the light, namely the gas, comes from one common reservoir. So the religious teachers of all climes and ages are but as many lamp-posts through which is emitted the light of the spirit flowing constantly from one source, the Lord Almighty.[15]

Not only is gas referred to, but the policeman and his lantern as well:

> A policeman can see with a dark lantern (bull's-eye) everyone upon whom he throws the rays, but no one can see him so long as he does not turn the light towards himself. So does God seeth everyone, but no one seeth Him until the Lord revealeth Himself to him in His mercy.[16]

Remarkable in its own way is the acceptance of Jesus Christ as a realized being and his incorporation into the traditional list of such people. Thus in Saying 60, Jesus is mentioned with such traditional figures as Suka Deva, Narada and Shankaracharya![17] Elsewhere, Christ's crucifixion is explained somewhat docetically thus:

> When the shell of an ordinary cocoa-nut is pierced through, the nail enters the kernel of the nut too. But in the case of the dry nut, the kernel becomes separate from the shell, and so when the shell is pierced the kernel is not touched. Jesus was like the dry nut, i.e. His inner soul was separate from His physical shell, and consequently the sufferings of the body did not effect Him.[18]

III

It is clear, then, that Folklore should not always be looked upon as impervious to modernity. Ramakrishna's Sayings indicate the possibility that even an articulate, if illiterate, mystic can absorb elements of modernity visible around him and thereby enrich Folklore.

References.

1. S.G.F.Brandon,ed., *A Dictionary of Comparative Religion* (New York: Charles Scribner's Sons, (1970) p.289.
2. ibid.
3. F. Max Mueller, *Ramakrishna: His Life and Sayings* (London : Longmans. Green & Co., 1898).
4. ibid., pp.29-30,96.
5. ibid., pp.95-96.
6. ibid., p.70.

7. ibid., p.42, emphasis added.
8. One should guard here against false attribution. For instance, Saying No.2 contains the word 'aqua' for water but it is highly unlikely that it would have been used as such by Ramakrishna. Obviously what R.R.K. Hartmann calls Contact Analysis is involved here. (See R.R.K. Hartmann, Talking on Con-Text, Hemisphere, Vol. 21, No9 [September 1977] p.16). Similarly associating Saying No.27 with the "Permanent Settlement of Bengal" may be a bit far-fetched, or Saying No.136 with Ramakrishna's actual knowledge of the Western musical scale.
9. F. Max Mueller, op.cit., p.103 (23).
10. ibid., p.165 (293).
11. ibid., pp.108-109(51,52).
12. ibid., p.108(48).
13. ibid., pp.176-177(336).
14. ibid., p.107(45).
15. ibid., p.148(224).
16. ibid., p.413(196).
17. ibid., p.111.
18. ibid., pp.111-112(62).

MAX MUELLER, RAMAKRISHNA, AND THE DIALOGUE OF WORLD RELIGIONS

I

It is generally believed that "As a result of the discoveries made in all branches of oriental studies during the first half of the nineteenth century, as well as of the establishment of Indo-European philology and comparative linguistics, the history of religions first really entered its own with Max Mueller (1823-1900)".[1] Thus launched, the science of religion or Comparative Religion[2] or the History of Religions,[3] gained academic recognition, and the "study of religion came to be regarded as an autonomous discipline."[4] A review of the recent literature and work in the field of the academic study of religion indicates a great deal of occupation, almost a preoccupation, with the phenomenon of what is called '*dialogue*'[5] among the religions of the world.

It is now over eighty years since Max Mueller died. This essay is an attempt to identify some thought-patterns of his which might be applicable to the current discussion. It is not an attempt to second-guess, only to guess what points he might have liked to stress in the current debate.

II

The first of these can be seen in the remarks which accompany his biography of Ramakrishna based on the material sent to him by Swami Vivekananda. In this context he refers to what he calls The Dialogic Process—which, by the way, has nothing to do with the 'dialogue' of world religions as understood today. (That it *may* have some implications for it will be shown later.) By the Dialogic Process Max Mueller seems to understand "the transformation which mere repetition, conversation, or what is called oral tradition will and must produce in the description of the facts as they really happened."[6]

He is then quick to draw attention to the fact that even Swami Vivekananda's account of his Master has undergone this process.

> I had made it as clear as possible to Vivekananda that the accounts hitherto published of his Master, however edifying they might be to his followers, would sound perfectly absurd to European students, that stories of miraculous events in childhood, of apparitions of goddesses (devi) communicating to the Sanyasin a knowledge of languages and literatures which we know he never possessed in real life, would simply be thrown away on us poor unbelievers, and that descriptions of miracles performed by the Saint, however well authenticated, would produce the very opposite effect of what they were intended for. Vivekananda himself is a man who knows England and America well, and perfectly understood what I meant. Yet even his unvarnished description of his Master discloses here and there the clear traces of what I call the Dialogic Process, and the irrepressible miraculising tendencies of devoted disciples. And I am really glad that it does so, if only it helps to teach us that no historian can ever pretend to do more than to show us what a man or a fact seemed to be to him or to the authorities whom he has to follow, and not where he or it actually was.[7]

Understood thus, the Dialogic Process is a historiographic phenomenon, which highlights the fact that "there is hardly a single fact in history which can escape being modified by this process before it reaches the writer of history."[8]

In the context of this discussion, however, Max Mueller throws up the idea of the Dialectic Process, which he "prefers to call by the wider name the *Dialogic Process.*"[9] This Diatectic Process,

> ....as applied to the facts of history comprehends all the changes which are inevitably produced by the mere communication and interchange of ideas, by the give and take of dialogue, by the turning of thoughts from one side to another. It is in reality what

> is called in German the threshing out of a subject, a kind of *Durchsprechen*, or what the Greeks call a speaking forward and backward, or dialogue.[10]

The conception of dialogue thus articulated by Max Mueller seems to be of some significance in the modern context of the world religions.

III

In the modern literature on the dialogue of world religions, various models have been developed about the kind of the religious world the present dialogue of world-religions is likely to usher in. There are three basic visions:(1) that a single religion will become a world religion;[11] (2) that the world will continue to be full of several religions[12] and (3) that a universal religion of some kind will emerge.[13] Within the second possibility there are a few sub-possibilities: (1) that all religions will become evangelical;[14] (2) that all religions will reconceive themselves in the light of the other religions;[15] (3) that all religions will tend to have more and more in common[16] and (4) that all religions will reaffirm themselves.[17]

In all of this discussion one fact tends to be ignored and it appears that Max Mueller's discussion of the term dialogue throws some light on this area. He seems to be pointing out that the "mere communication and interchange of ideas, the give and take of dialogue" is in itself an *altering* process. In other words, whatever dialogue may do to a religion or a religion to a dialogue, the mere decision to engage in dialogue irrespective of its goals or motives, is in itself consequential for a religion.

Is it because of this awareness that other religions of the world are taking their own time in responding to Western overtures for dialogue?

References

1 Mircea Eliade, *The Sacred and the Profane* (New York: Harcourt, Brace & World Inc.,1959) p.229.
2. S.G.F. Brandon, ed., *A Dictionary of Comparative Religion* (New York: Charles Scribner's Sons, 1970) p.202.
3. Mircea Eliade, *The Quest* (University of Chicago Press, 1969) Chapter I.
4. See Jacques Waardenburg, *Classical Approaches to the Study of Religion* (The Hague: Mouton, 1973) p.6.
5. See Edward J. Jurji, ed., *Religious Pluralism and World Community* (Leiden: E.J. Brill, 1969) passim; Robert Lawson Slater, *World Religions and World Community* (New York: Columbia University Press, 1963) passim; Ninian Smart, *A Dialogue of Religions* (London: SCM Press, 1960) passim: Herbert Jai Singh, ed., *Inter-religious Dialogue* (Bangalore: The Christian Institute for the Study of Religion, 1967) passim; J.G. Arapura, *Religion as Anxiety and Tranquility* (The Hague: Mouton, 1972) Chapter 9.
6. Max Mueller, *Ramakrishna: His Life and Sayings* (London: Longmans, Green & Co., 1898) p.25.
7. ibid., pp.29-30. Diacritics omitted.
8. ibid., p.25; also see pp.26-29, 60, 62.
9. ibid., p.25.
10. bid.
11. Robert Lawson Slater, op.cit., p.213.
12. Robert Lawson Slater, op.cit., p.217.
13. See Huston Smith, *The Religions of Man* (New York: Harper & Row, 1965) p.352.
14. But see E.G. Parrinder, *A Book of World Religions* (London: Hutton Educational Publications, 1965) p.172.
15. William Ernest Hocking, *The Coming World Civilization* (New York: Harper, 1956) passim.
16. George Rupp, *Christologies and Cultures* (The Hague : Mouton,1974) p.258.
17. Rudolf Otto is quoted as having remarked: "A gigantic struggle is developing. Its climax will, however, only be reached when mankind has achieved a certain equilibrium in social and political affairs. That will be the highest and holiest moment in the history of man. It will no longer be a question of political structures, economic groupings or social interests. For then the religions of mankind will arise against one another" (see Harold W.Turner, *The Idea of the Holy—A Guide for Students* [Aberdeen: Aberdeen People's Press,] 1974 p.7.).

21

RAMAKRISHNA PARAMAHAMSA: A STUDY IN A MYSTIC'S ATTITUDES TOWARDS WOMEN

I

Ramakrishna Paramahamsa is, as is well-known, one of the leading mystics of modern India.[1] He has left a deep impression on the attitude of modern Hindus towards other religions.[2] This aspect of his life and religious experience is quite well-known. This paper delves into a rather lesser known aspect of his life—his attitude towards women.

The attitude of a mystic towards women is, needless to say, of great spiritual and social significance, especially for a religion like Hinduism. Yet the study of the attitude of mystics towards women in Hinduism has been handicapped by the absence of first-hand accounts of their attitudes and behaviour towards women. For this reason alone, Ramakrishna deserves serious study, for his case provides a striking exception to this dearth of first-hand documentary material.[3] One of Ramakrishna's disciples, a certain Mahendranath Gupta, recorded the conversations of Ramakrishna along with the circumstances in which they took place "from March 1882 to April 24, 1886",[4] till only a few months before Ramakrishna's passing away. This diary, kept in Bengali, fills five volumes and has been available in English translation at least since 1942 under the title: "The Gospel of Sri Ramakrishna". This source, coupled with Swami Saradananda's well-known biography[5] of Ramakrishna called "Sri Ramakrishna, the Great Master", provides abundant material for a study of Ramakrishna's attitude towards women.

II

As one surveys this material one is struck by the fact that Ramakrishna's attitudes towards women do not seem to be, at least at first glance, entirely consistent. One might even say that he seems to have been ambivalent in his attitude towards women. Let us look at the evidence first.

In several contexts, Ramakrishna conveys a rather negative assessment of women. For Ramakrishna the purpose of life was God-realization. And what was his advice for those who wished to realize God?

> Those who wish to attain God or make progress in their devotional practices should particularly guard themselves against the snares of lust and wealth. Otherwise they will never attain perfection.[5]

The English words, lust and wealth, lack the colour of the original expression used by Ramakrishna-*Kamini-kancana* (woman and gold).[7] According to Ramakrishna, "the two obstacles to spiritual life are woman and gold."[8] And since he also said that "It is woman that creates the need for gold,"[9] we are in the end left really with only one obstacle to spiritual salvation—a woman. Thus, on one occasion, Ramakrishna told his followers and disciples:

> "Woman and gold" is the cause of bondage." "Woman and gold" alone constitutes *Samsara,* the world. It is "woman and gold" that keeps one from seeing God.

At this point he held a towel in front of his face and then asked his disciples:

> Do you see my face any more? Of course not. The towel hides it. No sooner is the covering of "woman and gold" removed than one attains *Chidananda,* Consciousness and Bliss.

Then he went on to say:

> Let me tell you something. He who has renounced the pleasure of a wife has verily renounced the pleasure of the world. God is very near to such a person.

The devotees listened to the words in silence.

> Master (to Kedar, Vijay and other devotees): He who has renounced the pleasure of a wife has verily renounced the pleasure of the world.
> It is "woman and gold" that hides God. You people have such imposing moustaches, and yet you too are involved in "woman and gold". Tell me if it isn't true. Search your heart and answer me.[10]

And then he went on to remark, "I see that all are under the control of woman."[11] Elsewhere he compares being under a woman's influence to being under a ghost's influence;[12] he also compares "woman and gold" to "the hog plum all stone and skin. If one eats it one suffers from colic."[13] He advises those who wish to develop true renunciation to "keep themselves at least fifty cubits away from women."[14] He states that "woman and gold', alone is the world: that alone is *maya.*"[15] He describes sitting with or talking to a woman for a long time as a "kind of sexual intercourse";[16] he testifies to the "bewitching power of women."[17] and claims that, "I have never enjoyed a woman, even in a dream."[18] Ramakrishna's marriage with Sarada Devi was never consummated.

It is clear from these statements that on several occasions, Ramakrishna shared the view, traditional in certain sections of Hinduism, that women constitute a major obstacle to God-realization.[19]

III

However, having said this, it would be unfair to Ramakrishna to stop here, for when one turns from these conversational snippets with his disciples to the biographical details of his life, the picture changes. The spiritual discipline or *sadhana* of Ramakrishna, which spanned twelve years, has been divided into three periods, from 1856-1859; from 1860-1863 and from 1864-1867.[20] In each of these three phases, "women" in some form played a key role in Ramakrishna's spiritual progress. It is well-known that in the first phase of *sadhana,* the major mystical experience of Ramakrishna consisted of a vision of *Kali*[21]—the Mother Goddess, an experience from which Ramakrịsna came out uttering "repeatedly the word 'Mother' in a plaintive voice."[22] Not only that, it took place in a temple which had been built by a low-class but rich widow.[23] In the second phase of his spiritual progress Ramakrishna was instructed by a lady called Bhairavi, also known as *Brahmani,* whose coming Ramakrishna had foreseen in a vision,[24] and who instructed him in *Tantrika* practices,[25] one of which involved the following procedure as described by Ramakrishna himself.

> On one occasion, I saw, that the Brahmani had brought at night—nobody knew whence-a beautiful woman in the prime of her youth, and said to me, "My child, worship her as the Devi". When the worship was finished, she said, "Sit on her lap, my child, and perform Japa". I was seized with fear, wept piteously and said to Mother, "O Mother, Mother of the universe, What is this command Thou givest to one who has taken absolute refuge in Thee? Has Thy weak child the power to be so impudently daring?" But as soon as I said so, I felt as if I was possessed by some unknown power and an extraordinary strength filled my heart. And uttering the *Mantras,* no sooner had I sat on the lap of the woman, like one hypnotized, unaware of what I was doing, than I merged completely in *Samadhi.* When I regained consciousness, I saw the Brahmani waiting on me and assiduously trying to bring me back to normal consciousness. She said, "The rite is completed, my child; others restrain themselves with very great difficulty under such circumstances and then finish the nominal *Japa* for trifling little time only; but you lost all consciousness, and were deep in *Samadhi*". When I heard this I became assured and began to salute Mother again and again.[26]

From the third phase of Ramakrishna's spiritual pilgrimage comes even more striking evidence of how "woman" can participate in spiritual life. Indeed, even womanhood itself may be embraced as a part of one's comprehensive spiritual seeking. This is known as *Madhura Bhava*[27] in *Bhakti sadhana*, in which the male devotee adopts the role of a female worshipper of God. When Ramakrishna adopted this mode, he was criticized for doing so but he disregarded the criticism. His biographer tells us that, adorned in feminine dress and ornaments—

> ...the Master gradually merged so much in the mood of the women of *Vraja* desirous to have the love of Krishna, that the consciousness that he was a male person disappeared altogether and every thought, word or movement of his became womanly . The Master, we were told by himself, was thus in a woman's dress for six months under the faith that he was the spiritual consort of God.[28]

There is some evidence to suggest that during this period Ramakrishna began to menstruate.[29]

Not only did women play an important role in the spiritual practices of Ramakrishna, but even after Ramakrishna had completed his *sadhana* and had won recognition as a spiritual figure, women continued to play a part in his life.[30]

It is clear, then, that Ramakrishna's attitude towards women is somewhat puzzling. On the one hand, woman is condemned as the greatest obstacle to God-realization; on the other, not only does he undergo a transvestite transfiguration himself to experience God, he sees God as inhabiting a woman's body. Can this ambivalence be reconciled?

IV

The situation does seem to this writer to be capable of reconciliation from a particular standpoint.[31] Although Ramakrishna's attitude to women is a variable and not a constant—it seems to vary predictably in relation to a central referent that of God—realization. Thus all his criticism is directed against women when they seem to present an obstacle to God-realization, as in the case of most men of the world; on the other hand, whenever woman stands in a positive relationship to God-realization, whether as an aspect of the Godhead itself, as *Kali,* or as one assisting in God-realization as a Teacher or guru, or as a devotional mode or as herself striving for God - realization, Ramakrishna has nothing against her.

The following conclusions can thus be drawn about Ramakrishna's attitude towards women:

(1) His attitude varies from time to time and is thus a variable;

(2) Although his attitude is variable, it is not arbitrarily so—it is not unpredictable;

(3) It is predictable in relation to a central referent, namely God-realization; and

(4) If women relate positively to the prospect of God-realization, his attitude towards them is positive[32] and if they relate negatively to the prospect of God-realization, his attitude is negative.

Or, to conclude, in Ramakrishna's own words:

As there are women endowed with *vidyasakti,* so also there are women with *avidyasakti*.A woman endowed with special attributes leads a man to God, but a woman who is the embodiment of delusion makes him forget God and drowns him in the ocean of worldliness.[33]

References

1. See R.C. Zaehner, ed., *The Concise Encyclopaedia of Living Faiths* (Boston: Beacon, 1959) p.257.
2. ibid.
3. See Aldous Huxley, "Foreword to Swami Nikhilananda, *The Gospel of Sri Ramakrishna* (New York: Ramakrishna-Vivekananda Center, 1942).
4. Swami Nikhilananda, op.cit., p.viii. Aldous Huxley describes him as Ramakrishna's Boswell! (p.v.).
5. Swami Saradananda (trans. Swami Jagadananda), *Sri Ramakrishna the Great Master* (Madras: Sri Ramakrishna Math, 1952).
6. *Sayings of Sri Ramakrishna* (Madras: Sri Ramakrishna Math, 1965) p.45.
7. ibid.,p.53,fn.I.
8. Swami Nikhilananda, op.cit., p.247.
9. ibid., p.166.
10. ibid., pp.438-439. Emphasis added, diacritics omitted. Ramakrishna sometimes drew a distinction between the householder and the *sanyasi* when it came to women. "One cannot obtain the knowledge of *Brahman* unless one is extremely cautious about women. Therefore it is very difficult for those who live in this world to get such knowledge. However clever you may be, you will stain your body if you live in a sooty room. The company of a young woman evokes lust even in a lustless man. But it is not so harmful for a householder who follows the path of knowledge to enjoy conjugal happiness with his own wife now and then. He may satisfy his sexual impulse like any other natural impulse. Yes, you may enjoy a sweetmeat once in a while". On the other hand sex "is extremely harmful for a *sannyasi*. He must not even look at the portrait of a woman. A monk enjoying a woman is like a man swallowing the spittle he has already spat out" (ibid., p.387).
11. ibid., p.439."Worldly people are in a state of chronic intoxication—mad with "woman and gold",—they are insensible to spiritual ideas. That is why I love the youngsters, not yet stained by "woman and gold". They are good receptacles and may be useful in God's work" (ibid., p. 781).
12. ibid
13. ibid., p.817.
14. ibid., p.603.
15. ibid., p.670. Emphasis added, diacritics omitted.
16. ibid., p.701.
17. ibid., p.748.

18. ibid., p.832. It must be kept in mind, however, that even in the course of this generally denigratory attitude, Ramakrishna makes exceptions. (1) He points out that one way to overcome passions is to be like a woman—a technique later suggested by Mahatma Gandhi as well. "How can a man conquer passion? He should assume the attitude of a woman. I spent many days as the handmaid of God. I dressed myself in woman's clothes, put on ornaments, and covered the upper part of my body with a scarf, just like a woman. With the scarf on I used to perform the evening worship before the image" (p.603). (2) He accepts that some women could lead man to God (p.216). (3) He regards all women as the "embodiments of Sakti" (p.336). "Again I saw a woman, wearing a blue garment, under a tree. She was a harlot. As I looked at her, instantly the ideal of Sita appeared before me. I forgot the existence of the harlot, but saw before me pure and spotless Sita, approaching Rama an Incarnation of divinity, and for a long time I remained motionless. I worshipped all women as representatives of the divine Mother. I realized the Mother of the Universe in every woman's form" *(Memoirs of Ramakrishna* ([Calcutta: Ramakrishna Vedanta Math, 1957] pp.172-173.)

19. It has been said that just as Ramakrishna warned men against women, he also warned women against men, that he "advised them to renounce lust and wealth and warned them against the snares of men" *(Life of Sri Ramakrishna* [Almora: Advaita Ashrama, 1924] p.524). It has also been pointed out that when Ramakrishna talks of 'woman and gold' "one must specifically bear in mind that the Master is propunding no cult of woman-hatred. The study of his sayings in their entirety... will show that his attitude towards womankind was one of profound respect bordering on worship. For he saw in them a symbol of the Divine Mother of the universe in a special sense. It was in this light that he viewed all women, and he wanted his devotees also to cultivate the same attitude. But at the same time he impressed on his disciples that, when viewed with a sensual eye, 'woman' was the greatest danger to the spiritual aspirant. Of course, by 'woman' he meant 'sex, or 'carnality,' but the great teacher that he was, with deep insight into the workings of the human mind, he preferred to use the conrete for the abstract, and always spoke of the bondage of 'woman' and not of 'sex'. His motive must have been purely psychological, for speaking on the same subject to his women devotees, he used to warn them against the dangers from 'man' *(Sayings of Sri Ramakrishna* [Madras : Sri Ramakrishna Math, 1965] pp.43,fn.I)

20. Swami Saradananda,op.cit.,p.153.
21. ibid., p.140.
22. ibid., p.141.
23. ibid., chap.IV.
24. ibid., pp.185-186.
25. ibid. chaps X,XI.
26. ibid., 195-196.
27. See Swami Nikhilananda, op. cit., p.115. Also see T.M.P. Mahadevan, *Outlines of Hinduism* (Bombay: Chetana, 1960) p.91.
28. Swami Saradananda op.cit., p. 234.
29. ibid., p.238.
30. See, *Life of Sri Ramakrishna* (Almora: Advaita Ashrama, 1924) pp.528, 539,etc.
31. There are at least two other ways in which some sort of a reconciliation between the two bodies of evidence presented earlier could be effected. (1) It could be effected by recognizing a discontinuity between the levels of the Master and the

disciples. Most of the criticism of "woman" is met with in the didactic material about Ramakrishna; on the other hand, their spiritual acceptance is reflected in the biographic material about his life. It could then be argued that as a Teacher he need have no scruples about mixing with women, but the disciples, being spiritually less advanced, had to be careful. There is some evidence to support this (see Swami Nikhilananda, op. cit., pp.595, 626, etc.) and there is a possible parallel here with that of the Buddha who advised Ananda, a disciple monk, not even to look at women but was himself visited by lay women devotees on some occasions at "odd hours". Nevertheless the reconciliatory power of this model is somewhat limited by the fact that even after Ramakrishna had become the Master he continued to be alert when it came to women. He said: "I am very much afraid of women. When I look at one I feel as if a tigress were coming to devour me. Besides, I find that their bodies, their limbs, and even their pores are very large. This makes me look upon them as she-monsters. I used to be much more afraid of women than I am at present. I wouldn't allow one to come near me. Now I persuade my mind in various ways to look upon women as forms of the Blissful Mother". He went on to say: "A woman is, no doubt, a part of the Divine Mother". But as far as man is concerned, especially *sanyasi* or a devotee of God, she is to be shunned. I don't allow a woman to sit near me very long, no matter how great her devotion may be. After a little while I say to her, 'Go and see the temples'. If that doesn't make her move, I myself leave the room on the pretext of smoking" (pp.593-594). Not only that, as a disciple he himself accepted instructions from a lady, hence this model only goes so far. (2) Another way to reconcile the two bodies of evidence would be to maintain that what the Master was opposed to was "sex" which for a man is symbolized by women and for a woman by men. But as most of his disciples were men, and those who kept records of his life and teachings only men, most of Ramakrishna's statements seem to be directed against women. This model again is helpful in understanding the situation somewhat but leaves unanswered the question: then why does he sometimes make positive statements about women (pp.336,425, 432) and why does he recommend adopting a woman's ways to overcome passions? (p.603).

32 Even sex, the great no-no of the spiritual path, is acceptable in the realm of God-realization, according to Ramakrishna. Thus when Ramakrishna was asked: "When one sees God, does one see him with these eyes", he said:

God cannot be seen with these physical eyes. In the course of spiritual discipline one gets a "love body", endowed with "love eyes", "love ears", and so on. One sees God with those "love eyes", One hears the voice of God with those "love ears". One even gets a sexual organ made of love.

At these words, M. burst out laughing. The Master continued, unannoyed, "While this 'love body' the soul communes with God".

M. again became serious" (Swami Nikhilananda, op.cit., p.115).

33. ibid.,p.216.

22

WHICH MYSTICAL EXPERIENCE OF SWAMI VIVEKANANDA WAS REALLY ADVAITIC?

Swami Vivekananda had several mystical experiences,[1] many of which were induced by Ramakrishna's touch.[2] This essay is an attempt to see which one of them, if any, may be properly called *advaitic*

I

One may begin by relating those experiences which were clearly not *advaitic*.[3] To this non-*advaitic* class would belong the semi-mystical experience of *Kali* which Swami Vivekananda had in Kashmir.

> While wandering as an itinerant *sanyasin* in Kashmir some time after his teacher had passed away, he came across an abandoned temple of *Kali*, evidently destroyed partially by a wave of invaders, but whose image was strangely beautiful. He stood at the entrance. Anger welled up and he thought how terrible it was that the Mother's temple should come to this. A voice from the image rang out, "My child, it is through my will that they are destroyed". He came away intoxicated with the vividness of the experience and a thoroughly changed outlook. Later, he said to sister Nivedita, referring to this experience, "There must be no more of this anger. Mother said, What even if the unbeliever should enter My temple and defile My images, what is that to you ? *Do you protect me? Or do I protect you?*"[4]

Earlier on, before Ramakrishna passed away, Swami Vivekananda had another mystical experience of *Kali* in a spiritual rather than a historical context.[5] Vivekananda went to the temple that night to pray to *Kali*. He relates

> As I was going, a sort of profound inebriation possessed me; I was reeling. A firm conviction gripped me that I should actually see Mother and hear her words. I forgot all other things, and became completely merged in that thought alone. Coming in the temple I saw that Mother was actually pure consciousness, was actually living and was really the fountain-head of infinite love and beauty. My heart swelled with loving devotion; and beside myself with bliss I made repeated salutations to Her, praying, "Mother, grant me divine knowledge and devotion, ordain that I may always have unobstructed vision of you". My heart was flooded with peace. The whole universe completely disappeared and Mother alone remained filling my heart.6

Although there is a reference to "Pure consciousness" in this passage, which is associated with the Brahman of Advaita Vedanta,[7] references to the form of the Mother and to devotion indicate that the experience was dualistic in nature.

It is clear, therefore, that Swami Vivekananda's experiences associated with the Mother cannot be regarded as *advaitic*.

II

There is, however, another experience which has been described as *advaitic* by several writers,[8] and perhaps even by Swami Vivekananda himself.9 On account of its importance, the report of the experience, though somewhat lengthy, needs to be cited in full:

> Thus one day the Master told Narendranath many things indicating the oneness of Jiva and Brahman of the non-dual philosophy. He heard those words, undoubtedly with attention, but he could not comprehend them and went to Hazra at the end of the Master's talk. Smoking and discussing those things again with Hazra, he said, 'Can it ever be possible that the waterpot is God, the cup is God, whatever we see and all of us are God? Sri Hazra also joined

Narendra in thus ridiculing the idea and both of them burst into laughter. The Master was till then in the state of partial external consciousness. Hearing Narendra laugh, he came out of his room like a boy with his wearing cloth in his armpit and, coming to them smiling, said affectionately, 'What are you both talking about?' He then touched Narendra and went into samadhi.

Narendra said to us afterwards, 'There was a complete revolution in the state of my mind in a moment at the wonderful touch of the Master. I was aghast to see actually that there was nothing in the whole universe except God. But I remained silent in spite of seeing it, wondering how long that state would last. But that inebriation did not at all diminish that day. I returned home; it was all the same there; it seemed to me that all that I saw was He. I sat for my meal when I saw that all—food, plate, the one who was serving as well as myself—were nothing but He. I took a mouthful or two and sat quiet. My mother's affectionate words-"why do you sit quiet; why don't you eat?"—brought me to consciousness and I began eating again. Thus, I had that experience at the time of eating or drinking, sitting or lying, going to the college or taking a stroll. I was always overwhelmed with a sort of indescribable intoxication. When I walked along the streets and saw a carriage coming along before me I did not feel inclined, as at other times, to move away, lest it should collide with me. For, I thought, "I am also that and nothing but that". My hands and feet always remained insensible at that time. I felt no satisfaction whatever when I took my food. It seemed to me as if some one else was eating the meal... When that overwhelming intoxication diminished a little, the world appeared to me to be a dream. Going for a walk on the bank of the Hedua tank, I knocked my head against the iron railings round it to see whether what I saw were dream-rails or actual ones. On account of the insensibility of my hands and feet I was afraid whether I was not going to have paralysis. I could not escape that terrible intoxicating mood and overwhelming condition for some time. When I came to the normal state, I thought that that was the indication of non-dual knowledge. So what is written in the *Sastras* about it is by no means untrue. Since then I could never doubt the truth of non-duality.[10]

Although this experience has been called *advaitic* or non-dual, the point needs to be examined. It is quite clear that the identity which Swami

Vivekananda felt *was between himself and the external world, not between himself and Brahman.* Thus the experience may be called non - dual in a general way, in the sense that he experienced non - duality between himself and the rest - of - the - world; but it cannot be called *advaitic* in the strict sense, for the *advaitic* experience consists of the realization of identity with *Brahman*, not the world.[11]

This mystical experience was clearly incomplete. Indeed in a sense it may be said to have been aborted and though it could have ended in a 'true' *advaitic* experience, it did not.[12] But the next experience to be examined seems to have been a full - fledged experience. During the last days of Ramakrishna's life Swami Vivekananda was often seen-

> ... urging the Master to allow him the highest superconscious revelation, the great ecstasy from which there is no return, the *Nirvikalpa Samadhi,* but Ramakrishna emphatically refused him. One day, Swami Shivananda told me, he was present in the garden at Cossipore, near Calcutta, when Naren[13] really attained this state. Seeing him unconscious, his body as cold as that of a corpse, we ran in great agitation to the Master and told him what had happened. The Master showed no anxiety; he merely smiled and said, 'Very well!' and then relapsed into silence. Naren returned to outward consciousness and came to the Master. The Master said to him, 'Well, now do you understand? This (the highest realization) will henceforth remain under lock and key. You have the Mother's work to do. When it is finished, She will undo the lock.'[14]

III

To conclude: it is only the last experience narrated above which seems to have been a truly *advaitic* mystical experience of Swami Vivekananda and not any of the others mentioned in the essay.[15]

References

1. For a useful account see Claude Alan Stark, *God of All* (Cape Cod, Mass: Claude Stark Inc., 1974) pp.123-137.

2. ibid., pp.125, 128, etc. For an interpretation of this touch "for scientific men" see Romain Rolland, *The Life of Ramakrishna* (Calcutta: Advaita Ashrama, 1960) p.240, fn.l. For trances induced by Ramakrishna's touch in other disciples, see Claude Alan Stark, op.cit., p.159, etc. On the lighter side one may add that the role of Ramakrishna's touch has led some to coin the expression *sparsa Yoga* as a new form of Yoga!
3. The order in which the experiences are narrated in this essay is not chronological.
4. Claude Alan Stark, op.cit., p.135. Second emphasis supplied.
5. For another mystical experience of a more general kind, see ibid., pp.132-133.
6. ibid., p.133.
7. See T.M.P. Mahadevan, *Outlines of Hinduism* (Bombay: Chetana Ltd., 1971) p.147.
8. See Claude Alan Stark, op.cit., p.128; Romain Rolland, op.cit., pp.239-240, etc.
9. See Swami Saradananda, *Sri Ramakrishna, The Great Master* (Madras: Ramakrishna Math, 1952), p.770.
10. ibid., pp.128-129. Emphasis supplied.
11. For the same reason Swami Yogananda's experience of not merely global but cosmic identity, though quite remarkable, cannot strictly be regarded as an *advaitic* experience; see Paramahamsa Yogananda, *Autobiography of a Yogi* (Los Angeles: Self-realization Fellowship, 1973) pp.165-168.
12. Another mystical experience induced by Ramakrishna sounds more like a divine debriefing than an *advaitic* experience (see Claude Alan Stark, op.cit., p.125).
13. Swami Vivekananda's former name, before he became a *Sanyasin*.
14. Romain Rolland, op.cit., p.246. Emphasis supplied. Also see Claude Alan Stark, op.cit., pp.136-137. It appears that early in his association with Ramakrishna, Vivekananda may have *come close* to having the *advaitic* experience, but was prevented in good time. It has been cited earlier. See Claude Alan Stark, op.cit., p.125.
15. According to Swami Balaramananda (personal communication, April 27, 1977) there are "five occasions when Swami Vivekananda is known to have entered *Nirvikalpa Samadhi*" and they are in brief:

    "1. At Hrishikesh, in 1890 (September) Swamiji had entered *Samadhi*. After regaining body-consciousness he said, to his brother-disciples, 'I had *Nirvikalpa Samadhi*' (Life of Swami Saradananda in Bengali, p.74).

    2. Prior to this at Almora in the August of 1890, he entered into a similar state. *(Life of Swami Vivekananda,* Vol.II, Mayavati: Advaita Ashrama, 1914) p.114.

    3. Sometime in 1894 on the shore of Lake Michigan at Chicago, he entered *Samadhi* (M.L. Burke, *Swami Vivekananda in America: New Discoveries,* [Mayavati: Advaita Ashrama, 1966] p.117).

    4. At Thousand Island Park, USA in the July or August of 1895 he entered into *Nirvikalpa Samadhi* (*Life of Swami Vivekananda* [Mayavati: Advaita Ashrama, 1965] p.368).

    5. At Camp Percy, New Hampshire, USA, in the June of 1895, Swamiji is known to have entered *Nirvikalpa Samadhi* (M.L. Burke, op.cit., p.687). These are the occasions traditionally known to us, when Swamiji had *Nirvikalpa Samadhi*".

THE ATTITUDE OF SWAMI VIVEKA-NANDA TOWARDS PRAYER

I

Prayer is one of the most common expressions of religious experience.[1] Although in some form or other it is to be found in almost all the religions of the world,[2] its expression is most commonly associated with the sentiment of love for or devotion to God.[3] In other words, it constitutes a key element in *Bhakti* Yoga or the Yoga of devotion.[4] As most Hindus practice *Bhakti* Yoga in some form or other,[5] prayer is perhaps the most important single element in the life of the Hindus.

This essay seeks to analyze the attitude of Swami Vivekananda towards this very popular expression of religiosity in Hinduism.

II

It is noteworthy that references to prayer in the writings of Swami Vivekananda are rather sparse.[6] References to *Karma* Yoga and *Jnana* Yoga also exceed references to Bhakti in his writings.[7] This comparative analysis suggests, therefore, that Swami Vivekananda was less articulate when it came to *Bhakti* Yoga than on matters relating to *Karma* Yoga and *Jnana* Yoga. Similarly, references to meditation and action far outweigh references to prayer in keeping with this pattern.[8]

An impressionistic analysis of the kind carried out above, however, is of a rough-and-ready sort and may also be affected by the manner of indexing. It is rather from a closer analysis of the references to prayer that we may hope to gain less superficial and more valid insights into the situation.

III

When these references to prayer are analyzed in the writings of Swami Vivekananda, one notices a curious phenomenon, which may be described as that of "squeezing". It seems that Swami Vivekananda tends to describe the "low" forms of prayer as a variety of *Karma* and the "high" form of prayer as an expression of *Advaita Vedanta*—so that prayer as traditionally understood virtually gets squeezed in between these two.

Swami Vivekananda regards praying to God for physical material comforts as a "low" form of prayer. He says:

> Every religion has prayers, but one thing you must bear in mind—praying for health or wealth is not *Bhakti*, it is *karma* or meritorious action. Praying for any physical gain is simply *Karma*, such as a prayer for going to heaven and so forth. One that wants to love god, to be a *Bhakta*, must discard all prayers. He who wants to enter the realm of light must first give up this buying and selling, this "shopkeeping" religion, and then enter the gates. It is not that you do not get what you pray for; you get everything but such praying is a beggar's religion.[9]

It is clear, then, that it is, firstly, not a very spiritual thing to pray for mundane things and secondly, that it is our own *Karma* which dispenses the results of our prayer and not some God. "Praying to God" is good *Karma* which produces good results; the good results come, not from God, but through the law of *Karma*. It is useful to note that this kind of *Karmic* situation is hardly consistent with *Karma* Yoga, for *Karma* Yoga involves "Work without Motive."[10] So when mundane prayers get answered, one is really not functioning at a spiritual plane at all. It is not *Bhakti* Yoga, and it is not *Karma* Yoga either-merely *Karma!*[11]

At the other end of the spectrum, Swami Vivekananda regards adoration of the Self as the true prayer. Thus he says:

> If this room is full of darkness for thousands of years and you come in and begin to weep and wail, "Oh the darknes", will the darkness vanish? Strike a match and light comes in a moment. What good will it do to think all your lives, "Oh, I have done evil, I have made many mistakes"? It requires no ghost to tell us that. Bring in the light and the evil goes in a moment. Build up your character and

manifest your real nature, the effulgent, the resplendent, the ever-pure and call It up in everyone that you see. I wish that everyone of us had come to such a state that even in the vilest of human beings we could see the Real Self within, and instead of condemning them, say "Rise thou effulgent one, rise thou who are always pure, rise thou birthless and deathless, rise almighty and manifest thy true nature. These little manifestations do not befit thee". *This is the highest prayer that the Advaita teaches.*[12]

This passage, though extended, has been quoted at length because of its multiple significance. It contains an implied criticism of *Bhakti* in its popular aspect of seeking succour from God as superstitious. The ghost-imagery is especially suggestive in this respect.[13] It suggests that prayer is a failure of nerve. Hence the call to resolute action and the identification of the *advaitic* invocation as the highest form of prayer.

It is clear, therefore, that Swami Vivekananda views prayer from the standpoint of his basic pro-*Jnana* Yoga [14] and pro-*Karma* Yoga orientation. It is not suggested hereby that Swami Vivekananda completely disregarded *Bhakti* Yoga.[15] Rather, it is suggested that he seems to attach relatively less weight to *Bhakti* Yoga vis-a-vis *Karma* Yoga and *Jnana* Yoga and that this attitude is reflected in the treatment of prayer in his writings.[16]

References

1. See S.G.F. Brandon, ed., *A Dictionary of Comparative Religion* (London: Weidenfeld & Nicolson, 1970) pp.507-508.
2. See James Hastings, ed., *Encyclopaedia of Religion and Ethics* (New York: Charles Scribner's Sons, 1919) pp.154-205.
3. See R.C. Zaehner, ed., *The Concise Encyclopaedia of Living Faiths* (Hutchinson of London, 1959) passim.
4. See A.L. Basham, *The Wonder That Was India* (London: Sidgwick and Jackson, 1956) pp.328-335.
5. See Ted J. Solomon, "Early Vaisnava Bhakti and Its Authochtonous Heritage" in *History of Religions*, August 1970, Vol.10, No.1, p.32.
6. See *The Complete Works of Swami Vivekananda* (fifth edition) (Calcutta: Advaita Ashrama, 1971) Vols I-VIII.

7. ibid., passim.
8. ibid., passim
9. *The Complete Works of Swami Vivekananda*, Vol.IV, p.38. Emphasis added. Also see ibid, Vol.VII, pp.82-84; Vol.V, pp.266-267, etc. In this passage Swami Vivekananda criticizes prayer not because it is not answered but because it reflects an inferior mentality. However, he also criticizes prayer because it does not seem to be answered by him towards whom it is directed. "Tell me one case where these prayers have been answered. All the answers that came were from your own hearts" (ibid., Vol.II, p.303). Elsewhere he remarks: "But help does not come from without, it comes from within ourselves. Cry to all the gods in the universe, I cried for years, and in the end I found that I was helped. But help came from within" (ibid., p.356).
10. ibid., vol.V, p.246.
11. Swami Vivekananda makes an interesting comment on prayer in relation to *karma Kanda* (ritual). He observes that in the Mimamsa tradition, with its emphasis on ritual "What in other religions is called prayer disappeared and the Vedas became the gods" (*The Complete Works of Swami Vivekananda*, Vol.I, p.447).
12. ibid., Vol.II, p.347, emphasis added.
13. "You know in your inmost heart that many of your limited ideas, this humbling of yourself and praying and weeping to imaginary beings are superstitions... You know there are no ghosts, but no sooner are you in the dark than you feel a little creepy sensation" (ibid., Vol.II, p.303).
14. At some places he accepts the usefulness of prayer as an aid to meditation, vide ibid., Vol.IV, p.228, etc. He also remarks: "Prayers, ceremonials, and all other forms of worship are simply kindergartens of meditation" (ibid., p.249).
15. See ibid., Vol.IV, pp.3-51.
16. This attitude of Swami Vivekananda is so pervasive that he interprets prayer even in a Christian context with an *Advaitic* preference. "Note the Common Prayer...: 'Our Father which art in Heaven, halloweed be Thy name', and so on—a simple prayer, a child's prayer. Mark you, it is the 'Common Prayer' because it is intended for the uneducated masses. To a higher circle, to those who had advanced a little more, he gave a more elevated teaching: 'I am in my Father, and ye in me, and I in You'... And then, when the Jews asked him who he was, he declared that he and his Father are one..." (ibid., Vol.IV, p.148).

24

NEW MYTHS IN HINDUISM: THE SARASVATI MYTH IN THE LIFE OF RAMAKRISHNA AND VIVEKANANDA

I

Sarasvati is the goddess of learning in the Hindu pantheon.[1] In the case of both Ramakrishna and Vivekananda, a body of mythical lore has arisen associating this goddess with them. The purpose of this essay is both to explore *how* this has happened and *why* it has happened.

II

The first well-known biography of Ramakrishna is by Max Mueller,[2] based on the material supplied to him by Swami Vivekananda himself.[3] It seems that already, by this time, the myth had grown that Ramakrishna had been instructed in Hindu lore by the goddess Sarasvati or an incarnation thereof.

In requesting for an account of his Master's life from Swami Vivekananda, Max Mueller tells us, he 'made it as clear as possible to Vivekananda that the accounts hitherto published of his Master, however edifying they might be to his followers, would sound perfectly absurd to European students, that stories of miraculous events in childhood, *of apparitions of goddesses (devi) communicating to the Sanyasin a knowledge of languages and literature which, as we know, he never possessed in real life* would simply be thrown away on poor unbelievers...'[4]

Vivekananda's account, as related by Max Mueller, does refer to a lady of great learning. It seems that at one time during his austerities Ramakrishna felt a burning sensation all over his body:

I used to stand in the waters of the Ganges, with my body immersed up to the shoulders and a wet towel over my head all through the day, for it was insufferable. Then a *Brahman* lady came and cured me of it in three days. She smeared my body with sandalwood paste and put garlands on my neck, and the pain vanished in three days. Now this *Brahman* lady was, we are told, an extraordinary Bengali woman. She was versed in the philosophies and mythologies of India, and could recite book after book from memory. She could hold her ground in argument with the best pandits of the country. Tall and graceful, she combined in herself all the physical and intellectual qualities that would raise any man or woman high above ordinary mortals. She had a fine voice and was well versed in music. She had given up the world, practised Yoga (ascetics), attained to some wonderful Yogic powers, and was roaming all over India in the red garb of a Sanyasin. Nobody knew anything of her birth or family or name even, and nobody could induce her to say anything about them. She was as if some goddess had come to this earth to help men to perfection, moved by the sorrows and sins of this wicked world. She seemed to have known full well that she was destined to help three particular personages, who were very advanced in attaining perfection. Ramakrishna had been informed by his divine mother that she would come and teach him certain way to attain perfection. He recognised her at once, and she recognised him and said, 'I have found out the other two, and have been searching for thee for a long, long time, and today I have found thee'. Up to this time Ramakrishna had not found a single soul who could understand his superhuman devotion and perfect purity, and the arrival of this woman was therefore a great relief to him. His devotion and love knew no bounds.

All people were astonished at the wonderful learning of this Brahman lady, but they could not understand how she could sympathise and place even above herself this half-crazed Ramakrishna they took him for. To prove that she was not mad, the lady mentioned some Vaishnava scriptures, got the manuscripts from some learned pandits, and quoted passage after passage, showing that all these physical manifestations come to an ardent lover of God. It was recorded in these books that all these states physical and mental did happen to the great religious reformer of Bengal, Sri Chaitanya, four hundred years back, and the remedies

> were given, too, by which he overcame them. For instance, this burning sensation, as if all the body were in flames, from which Sri Ramakrishna was suffering at the time, was mentioned in these Vaishnava scriptures as having happened to the sheperdess of Braja, to the stainless Sri Radha, the beloved of Krishna, centuries before, and again in later times to Sri Chaitanya, when both of them felt deeply the pain of separation from their beloved (God). In both these cases relief came by smearing the body with sandalwood paste and wearing garlands of sweet-scented flowers. The lady held it to be no real disease, but a state of physical disturbance, which would would come to all who arrive at that stage of Bhakti, or love of God. She applied the same remedies for three days, and the trouble passed away.
>
> At another time during her stay he suffered much from an insatiable appetite. However much he might eat, the appetite was there, preying upon him as if he had taken nothing. The Brahman lady assured him that the same had happened to Chaitanya and other Yogins, and ordered all sorts of dishes to be put into his room on every side, day and night. This practice was continued for a few days, and the sight of so much food gradually acted upon the mind, and the false sensation passed away.
>
> The lady lived there for some years, and made her friend practise all the different sorts of Yoga which make a man complete master of his body and mind, render his passions subservient to his reason, and produce a thorough and deep concentration of thought, and, above all, the fearless and unbiased disposition which is essential to everybody who desires to know the truth and the whole truth.[5]

Remarking on the account of *Brahmana* lady, Max Mueller observes that the "story, for instance, of the *Brahmana* lady who was sent as a messenger and teacher to Ramakrishna, will sound to us far more probable. But when I first heard of it, this lady was represented as a kind of goddess who met her pupil in a forest and instructed him, like another Sarasvati, in all the Vedas, Puranas and philosophies. The difficulty that had to be solved by this heavenly apparition was, no doubt, the fact that Ramakrishna had never received a proper classical education, and yet spoke with authority about the ancient literature and religion to his countrymen."[6]

One sees perhaps at work here what Max Mueller elsewhere in the book refers to as the *Mythological Process*.[7] Most of the ancient sacred literature of India is in Sanskrit and the fact is that Ramakrishna was "ignorant of Sanskrit, nay, ... [that] he did not know a single word of the sacred language of India, is denied by nobody, and has been distinctly asserted by one of his great admirers, the Rev. P.C. Mozoomdar."[8] Hence the *dea ex machina.*[9]

One may now turn to the consideration of a similar situation with regard to his disciple Vivekananda. In the case of Ramakrishna the problem was that he was illiterate and yet served as a spokesman of the tradition. With Vivekananda the problem was not that he was illiterate but that he had had a modern Western education and not a classical Hindu one. He had corresponded, not with the Pundits of Benaras, but with Spencer.[10] 'According to Saradananda, Naren pursued the study of Western philosophy between his first examination in 1881 and that of 1884, corresponding to our licentiate's degree'.[11] It is clear, then, that Swami Vivekananda's formal education was in Western philosophy and acquired through English.[12] In 1893, however, he stood before the parliament of Religions as the spokesman of Hinduism, a role for which he would properly have to be equipped with a background in Sanskrit education which seems to have been wanting. The scene is best described in a letter written by Swami Vivekananda on November, 2, 1893 from Chicago. The incidents described occurred on September 11, 1893:

> There was a grand procession, and we were all marshalled on to the platform. Imagine a hall below and a huge gallery above, packed with six or seven thousand men and women representing the best culture of the country, and on the platform learned men of all the nations of the earth. And I, who never spoke in public in my life, to address this august assemblage! It was opened in great form with music and ceremony and speeches; then the delegates were introduced one by one, and they stepped up and spoke. Of course my heart was fluttering and my tongue nearly dried up; I was so nervous, and could not venture to speak in the morning. Mazoomdar made a nice speech, Chakravarti a nicer one, and they were much applauded. They were all prepared and came with ready-made speeches. I was a fool and had none, *but bowed down to Devi Sarasvati and stepped up,* and Dr Barrows introduced me. I made a short speech. I addressed the assembly as "Sisters and Brothers of America", a deafening applause of two minutes followed, and then I proceeded; and when it was finished I sat down, almost exhausted

> with emotion. The next day all the papers announced that my speech was the hit of the day, and I became known to the whole of America. Truly has it been said by the great commentator Shridhara ..... "Who maketh the dumb a fluent speaker". His name be praised! From that day I became a celebrity, and the day I read my paper on Hinduism, the hall was packed as it had never been before. I quote to you from one of the papers: "Ladies, ladies, ladies packing every place—filling every corner, they patiently waited and waited while the papers that separated them from Vivekananda were read", etc. You would be astonished if I sent over to you the newspaper cuttings, but you already know that I am a hater of celebrity. Suffice it to say, that whenever I went on the platform, a deafening applause would be raised for me. Nearly all the papers paid high tributes to me, and even the most bigoted had to admit that "This man with his handsome face and magnetic presence and wonderful oratory is the most prominent figure in the parliament", etc. etc. Sufficient for you to know that never before did an Oriental make such an impression on American society.[13]

It seems from the letter as if Swami Vivekananda had recast the series of events at Chicago and those leading up to it, perhaps unconsciously, in a divine mould. Thus the hand of God is seen in the chance meeting with Dr Wright, Professor of Greek at Harvard University, which ultimately enabled Swami Vivekananda to participate in the Conference.[14] In the same letter the Swami comments later on: "Day by day I am feeling that the Lord is with me, and I am trying to follow his direction. His will be done."[15]

The mythological process is again at work, it seems, this time not in the accounts of the Master by others, but of the disciple by himself. The Sarasvati myth in the context of the Ramakrishna mission is an interesting illustration of how new myths emerge even in a religion as old as Hinduism—and for a good reason. In the cases of Ramakrishna and Vivekananda the Sarasvati myth serves to fill the gap, so to say, caused by their lack of traditional formal learning. The Sarasvati myth seeks to blend tradition with modernity; or even to appropriate modernity in terms of tradition and thus come to terms with it.

References

1. See A.A. Macdonell, *The Vedic Mythology* (Delhi: Indological Book House, 1971) pp.86-88. A.L. Basham, *The Wonder That Was India* (New York: Taplinger Publishing Co., 1967) pp.318-319; Benjamin Walker, *The Hindu World,* Vol.I (New York: Frederick A. Praeger, 1968) pp.165-166, etc.
2. See Romain Rolland, *The Life of Ramakrishna* (Calcutta: Advaita Ashrama 1960) pp.297-302; Christopher Isherwood, *Ramakrishna and His Disciples* (London: Methuen & Co. Ltd., 1965) pp.335-337, etc.
3. F. Max Mueller, *Ramakrishna: His Life and Sayings* (London: Longmans, Green & Co., 1898) pp.24, 59-60.
4. ibid., p.29, emphasis added. diacritics omitted.
5. ibid., pp.43-45. Diacritics omitted.
6. ibid., p.60.
7. ibid., p.25.
8. ibid., pp.60-61.
9. Which, Max Mueller feels, was really not required.
10. Romain Rolland, op.cit., p.225.
11. ibid., fn.2 Romain Rolland points out that he "had then read Descartes, Hume, Kant, Fichte, Spinoza, Hegel, Schopenhauer, Auguste Comte and Darwin. But it seems to me that he can only have read them superficially from general treatises and that he did not study their actual works. He also followed a course of medicine, studying the physiology of the brain and nervous system" (ibid).
12. This does not mean that Vivekananda knew no Sanskrit at all but rather that his expertize lay in English while his familiarity with Sanskrit was that of an average literate Bengali.
13. *The Complete Works of Swami Vivekananda* Vol.V (Calcutta: Advaita Ashrama, 1959 [Mayavati Memorial edition] pp.20-21, emphasis added). For other references to Sarasvati see Sailendra Nath Dhar, *A Comprehensive Biography of Swami Vivekananda* (Part I) (Madras: Vivekananda Prakashan Kendra, 1975) p.461, etc.
14. ibid., p.2
15. ibid., p.23

25

RAMAKRISHNA AND THE SPIRITUAL POVERTY OF HISTORICISM IN HINDU STUDIES

I

Modern studies in Hinduism, especially academic, have been largely historical.[1] Indeed, this historical element in Hindu studies has been so marked that a Hindu scholar has been constrained to remark:

> ...there is no doubt that the Hindu writers went to one extreme in ignoring history altogether[2] while modern Western writers go to the other extreme in making too much of historical treatment of thought and art and digging at the roots of the tree instead of enjoying its flower and fruit.[3]

This tendency to historicize everything often deprives one of deeper spiritual insights into the phenomenon being studied. As a matter of fact, it may even be that "The truly religious mind does not care for the history of religion, just as, among us, the scientific mind does not dwell on the history of science."[4]

In order to see this contrast between the historical and the spiritual interpretation of a religious fact, it is interesting to take up two instances and see how a historian and a man of religion deal with them. A historian's view is easily obtained because there are so many of them, but where do we find a man of religion? Fortunately Ramakrishna Paramahamsa-a man of religion if ever there was one, has expressed himself on some points on which the historians too have presented their views. This seems to have happened in at least two cases; (1) the reference to the cows in the *RgVeda* and (2) the dark colour of the god Krishna and the goddess Kali.

The frequent references to the bull, the cow and the calf in the *RgVeda* has led historians to state that in the early Vedic Age "the rearing of cattle and other domestic animals was scarcely less important than agriculture. Cows were held in much esteem and milk ... formed an important part of the diet in the Vedic household. Herds of cattle were daily led to the pasture by *gopa* (cowherd). The valley of the *Yamuna* was especially famous for its wealth of kine."[5]

All this is fine. But one should not forget the basic fact that the *RgVeda* is a religious text. Many of the references to kine are also in that context. Thus *RgVeda* II.2.2 declares-they call upon you, O *Agni*, day and night like the cows to the calf. And *RgVeda* VI.45.28 states that these hymns call on you as the milchcows to the calf. What are we to make of this, apart from concluding that the Aryans drank milk?

A clue is provided to the spiritual significance of these statements by the following piece of conversation between Ramakrishna and a pundit which seems to provide an insight which eludes the historian:

> Master: ... Yes, the way to realize God is through discrimination, renunciation and yearning for Him. What kind of yearning? One should yearn for God as the cow, with yearning heart, runs after its calf.
> Pundit: The same thing is said in the Vedas: 'O God, we call on Thee as the cow lows for the calf'.
> Master: Add tears to your yearning...[6]

The dark colour of Krishna and Kali has attracted the attention of historians. Thus it has been observed that "the name means 'black', and the god is usually depicted as of that colour. Perhaps the oldest clear reference to the pastoral Krishna is in early Tamil anthologies, where 'the Black One' (*Mayon)* plays his flute and sports with milkmaids. He may have been originally a fertility god of the Peninsula."[7] He is thus regarded as a Dravidian or at least a pre-Aryan god.[8] Similarly, Kali is regarded as an aboriginal goddess at least in part because of her dark complexion.[9]

This approach to Krishna and Kali which emphasizes their dark complexion has led historians into drawing certain conclusions or making certain surmises. But there again, it seems, there is something which eludes the historian—the possible spiritual as distinguished from the historical significance of the dark complexion. Ramakrishna sheds light on this aspect when he says:

> Why have the images of Krishna and Kali, a dark-blue colour? That too is on account of distance, like the water of a lake, which appears green, blue, or black from a distance. Go near, take the water in the palm of your hand, and you will find that it has no colour. The sky also appears blue from a distance. Go near and you will see that it has no colour at all.
> Therefore I say that in the light of *Vedantic* reasoning, *Brahman* has no attributes. The real nature of *Brahman* cannot be described. But so long as your individuality is real, the world also is real, and equally real are the different forms of God and the feeling that God is a Person.[10]

IV

It is clear then that there are certain religious insights which elude a historian of religion. This does not mean that there should be no history of religion for "the historical treatment of a religion has its value even for the most religious men. For instance, it is useful for us to know what doctrines of Hinduism belong to the Vedic period and what doctrines belong to the later Puranic period, and what causes contributed to the growth and vigour of Hinduism in one period and what causes brought about its decay in another."[11] But what this discussion does demonstrate is that we should be on our guard against becoming so obsessed with the historical aspect of the religious that we end up by ignoring the religious aspect of the historical. It is always useful, especially in religious studies, to be reminded of a certain spiritual poverty of historicism.

References

1. See Arthur A. Macdonell, *A History of Sanskrit Literature*. (New York: Haskell House Publishers, 1968) Chapter I; K.M. Panikkar, *The Foundations of New India* (London: George Allen & Unwin, 1963) pp.68-69, etc.

2. This may be an exaggeration, see K. Satchidananda Murty, ed., *Readings in Indian History, Politics and Philosophy* (London: George Allen & Unwin, 1967) Part IA; A.K. Warder, *An Introduction to Indian Historiography* (Bombay: Popular Prakashan, 1972) passim.
3. D.S. Sarma, *Hinduism Through the Ages* (Bombay: Bharatiya Vidya Bhavan, 1961) p.l.
4. Sir Charles Eliot, quoted by D.S. Sarma, ibid.
5. R.C. Majumdar. H.C. Raychaudri & Kalikinkar Datta, *An Advanced History of India* Bombay: Macmillan India, 1974) p.33. Diacritics omitted.
6. Swami Nikhilananda, *The Gospel of Sri Ramakrishna* (New York: Ramakrishna-Vivekananda Center, 1942) p.485.
7. A.L. Basham, *The Wonder that Was India* (London: Sidgwick & Jackson, 1956) p.305. Diacritics omitted; spelling modified, emphasis added.
8. See Benjamin Walker, *The Hindu World*, Vol.I (London: George Allen & Unwin, 1968) p.559ff.
9. ibid., p.509.
10. Swami Nikhilananda, op.cit., p.150. Emphasis added.
11. D.S. Sarma, op.cit., p.l.

BIBLIOGRAPHY

Swami Abhedananda (Compiler), *Memoirs of Ramakrishna* (Calcutta: Ramakrishna Vedanta Math, 1957). See under Memoirs of Ramakrishna).✓

C.N.K. Aiyer, *Life and Times of Sankara* (Madras, 1934).

J.G. Arapura, *Religion as Anxiety and Tranquility* (The Hague: Mouton, 1972).

A.L. Basham, "Hinduism" in R.C. Zaehner (ed.) *The Concise Encyclopaedia of Living Faiths* (Boston: Beacon Press, 1959).

———, *The Wonder That Was India* (London: Sidgwick and Jackson, 1956). (New York: Taplinger Publishing Co., 1967).

Bhagavadgita

Krishna Datta Bharadwaj, *The Philosophy of Ramanuja* (New Delhi: Sir Shankar Lall Charitable Trust Society, 1958).

Agehananda Bharati, *The Ochre Robe* (London: George Allen & Unwin, 1961).

Haridas Bhattacharyya (ed.), *The Cultural Heritage of India* (Calcutta: The Ramakrishna Institute of Culture, 1956).

A.C. Bouquet, *Hinduism* (New York: Hutchinson's University Library, 1948).

S.G.F. Brandon (ed.), *A Dictionary of Comparative Religion* (New York: Charles Scribner's Sons, 1970).

M.L. Burke, *Swami Vivekananda in America: New Discoveries* (Mayavati: Advaita Ashram, 1966.)

Emily Caillet, *The Clue to Pascal* (London: S.C.M. Press, 1944).

Joseph Campbell (ed.), Heinrich Zimmer: *Philosophies of India* (London: Routledge & Kegan Paul, 1953).

John Craisted Carman, *The Theology of Ramanuja* (Yale University Press, 1974).

Bhagavan Das, *The Essential Unity of All Religions*, (Benares: The Indian Bookshop, 1947).

S.N. Dasgupta, *Hindu Mysticism* (New York: Frederick Ungar Publishing Co., 1959).

———, *A History of Indian Philosophy* (Cambridge University Press, 1963) Vols I & II.

Wm. Theodore de Bary (ed.), *Sources of Indian Tradition* (New York: Columbia University Press, 1958).

Amaury de Riencourt, *The Soul of India* (London: Jonathan Cape, 1961).

Eliot Deutsch & J.A.B. van Buitenen, *A Source Book of Advaita Vedanta* (Honolulu: University of Hawaii Press, 1971).

Nalini Devdas, *Sri Ramakrishna* (Bangalore: The Christian Institute for the Study of Religion and Society, 1966).

Sailendra Nath Dhar, *A Comprehensive Biography of Swami Vivekananda* (Madras: Vivekananda Prakashan Kendra, 1975).

Mariasusai Dhavamony, *Love of God According to Saiva Siddhanta* (Oxford: Clarendon Press, 1971).

Heinrich Dumoulin, *A History of Zen Buddhism* (Boston: Beacon Press, 1963).

Mircea Eliade, *Patterns in Comparative Religion* (London: Sheed and Ward, 1958).

————, *The Quest* (University of Chicago Press, 1969).

————,*The Sacred and the Profane* (New York: Harcourt Brace & World Inc., 1959).

————, *Yoga* (New York: Pantheon Books, 1954).

Ainslie T. Embree (ed.), *The Hindu Tradition* (New York: Vintage Books, 1972).

Encyclopaedia Britannica (Chicago: William Benton, 1961).

J.N. Farquhar, *Modern Religious Movements in India* (New York: Macmillan Co., 1915).

Harold W. French, *The Swan's Wide Waters* (London: Kennikat Press, 1974).

M.K. Gandhi, *All Men Are Brothers* (Paris: UNESCO, 1959).

———, *Hindu Dharma* (Ahmedabad: Navajivan Publishing House, 1958).

———, *The Story of My Experiments with Truth* (Washington DC: Public Affairs Press, 1940).

Swami Ghanananda, *Sri Ramakrishna and His Unique Message* (London: The Ramakrishna Vedanta Centre, 1970).

R.R.K. Hartmann, "Talking on Con-Text", *Hemisphere,* Vol.21, No.9 (Sept.1977).

James Hastings (ed.) *Encyclopaedia of Religion and Ethics* (New York: Charles Scribner's Sons, 1919).

S.S. Hasurkara, *Vacaspati Misra on Advaita Vedanta* (Darbhanga: Mithila Institute, 1958).

M. Hiriyanna, *Outlines of Indian Philosophy* (London: George Allen & Unwin, 1932).

William Ernest Hocking, *The Coming World Civilization* (New York: Harper, 1956).

Thomas J. Hopkins, *The Hindu Religious Tradition* (Belmont CA: Dickenson Publishing Co., 1971).

Aldous Huxley, Foreword to Swami Nikhilananda (tr.), *The Gospel of Sri Ramakrishna* (New York: Ramakrishna-Vivekananda Centre, 1942).

Christopher Isherwood, *Ramakrishna and His Disciples* (London: Methuen & Co. Ltd., 1965).

Swami Jagananda (tr.), *Sri Ramakrishna The Great Master* (by Swami Saradananda) Madras: Sri Ramakrishna Math, 1963). (Also under "Swami Saradananda").

William James, *The Varieties of Religious Experience* (New York: The Modern Library, 1929).

Edward J. Jurji (ed.), *Religious Pluralism and World Community* (Leiden: E.J. Brill, 1969).

J.B. Kripalani, *Gandhi: His Life and Thought* (Government of India: Publications Division, 1971).

Alexander Lipski, "Some Methodological Problems Encountered by Western Scholars in the Study of Modern Indian Religious History", *Journal of Indian History* Vol.43, 1965).

Arthur A. Macdonnell, *A History of Sanskrit Literature* (New York: Haskell House Publishers, 1968).

———, *The Vedic Mythology* (Delhi: Indological Book House, 1971).

T.M.P. Mahadevan, *The Hymns of Sankara* (Madras: Ganesh & Co., 1970).

———, *Outlines of Hinduism* (Bombay: Chetana Ltd., 1956 [1960 and 1971]).

——, *The Philosophy of Advaita with Special Reference to Bharatitirtha Vidyaranya* (Madras: Ganesh & Co., 1957).

T.F. Mainkar, *Mysticism in the RgVeda* (Bombay: Popular Books Department, 1961).

R.C. Majumdar, H.C. Raychaudhuri & Kalinkindar Datta, *An Advanced History of India* (London: Macmillan & Co., 1950 [Bombay: Macmillan India, 1974]).

G.P. Malalasekera (ed.), *Encyclopaedia of Buddhism* (Government of Ceylon, 1966).

Robert A. McDermott (ed.), *Basic Writings of S. Radhakrishnan* (Bombay: Jaico Publishing House, 1970).

Memoirs of Ramakrishna, (Calcutta: Ramakrishna Vedanta Math, 1957 [Also under Swami Abhedananda, compiler]).

F. Max Mueller, *Ramakrishna: His Life and Sayings* (London: Longmans, Green & Co., 1898).

Mundakopanisad

K.Sakchidananda Murty (ed.), *Readings in Indian History, Politics and Philosophy* (London: George Allen & Unwin, 1967).

P. Mutharasu, *The Life of Sri Ramalingar* (Tirunellveli: Samarasa Sanmarga Sangram, 1961).

V.S. Narvane, *Modern Indian Thought* (New York: Asia Publishing House, 1964).

Walter G. Neevel Jr., "The Transformation of Sri Ramakrishna" in Bardwell L. Smith (ed.), *Hinduism: New Essays in the History of Religions* (Leiden: E.J. Brill, 1976).

The New Encyclopaedia Britannica, Macropaedia, Vol.12 (Chicago: William Benton, 1974)

Swami Nikhilananda (tr.), *The Gospel of Sri Ramakrishna* (New York: Ramakrishna-Vivekananda Centre, 1942 [Madras: Sri Ramakrishna Math, 1964]).

John B. Noss, *Man's Religions* (New York: The Macmillan Co., 1967).

Arthur Osborne, *Buddhism and Christianity in the Light of Hinduism* (London: Rider & Co., 1959).

Rudolph Otto, *Mysticism East and West* (Toronto: The Macmillan Co., 1960).

K.M. Panikkar, *The Foundations of New India* (London: George Allen & Unwin, 1963).

E.G. Parrinder, *A Book of World Religions* (London: Hutton Education Publications, 1965).

Geoffrey Parrinder, *Avatar and Incarnation* (London: Faber and Faber, 1970).

Karl H. Potter, *Bibliography of Indian Philosophies* (Delhi: Motilal Banarsidas, 1970).

Michael Pye, *Comparative Religion* (Devon: David & Charles, 1972).

S. Radhakrishnan, *Eastern Religions and Western Thought* (Oxford University Press, 1940).

———, *The Hindu Way of Life* (London: Unwin Books, 1975; first published 1927).
———, *Indian Philosophy* (London: George Allen & Unwin, 1951 [1971]).
———, *The Vedanta According to Sankara and Ramakrishna* (London: George Allen & Unwin, 1928).
P.T. Raju, *The Philosophical Traditions of India* (University of Pittsburgh Press, 1971).
Swami Ramakrishnananda, *Life of Sri Ramakrishna* (Madras: Sri Ramakrishna Math, 1959).
N.V. Raman, *Sankaracarya the Great* (Madras, 1923).
Richard H. Robinson, *The Buddhist Religion: A Historical Introduction* (Belmont, CA: Dickenson Publishing Co. Inc., 1970).
Romain Rolland, *The Life of Ramakrishna* (Almora: Advaita Ashram, 1931 [Calcutta: Advaita Ashrama, 1960]).
George Rupp, *Christologies and Cultures* (The Hague: Mouton, 1974).
Swami Saradananda, *Sri Ramakrishna The Great Master* (Madras: Sri Ramakrishna Math, 1951 [also under Swami Jagananda, tr.]).
D.S. Sarma, *Hinduism Through the Ages* (Bombay: Bharatiya Vidya Bhavan, 1961).
———, *Studies in the Renaissance of Hinduism in the 19th and 20th Centuries* (Benares Hindu University, 1944).
Sayings of Sri Ramakrishna, (Madras: Sri Ramakrishna Math, 1965).
Ben-Ami Scharfstein, *Mystical Experience* (Oxford: Basil Blackwell, 1973).
Gerschom G. Scholem, *Major Trends in Jewish Mysticism* (New York: Schocken Books, 1974).
Frithjof Schuon, *Logic and Transcendence* (New York: Harper & Row, 1975).
K.M. Sen, *Hinduism* (Baltimore: Penguin, 1973 [first published 1961]).
Herbert J. Singh (ed.), *Inter-religious Dialogue* (Bangalore: The Christian Institute for the Study of Religion, 1967).
Robert Lawson Slater, *World Religions and World Community* (New York: Columbia University Press, 1963).
Ninian Smart, *A Dialogue of Religions* (London: SCM Press, 1960).
———, "Interpretation and Mystical Experience" *Religious Studies*, Vols 1-2, 1966-67.
Huston Smith, *The Religions of Man* (New York: Harper & Row, 1958 [1965]).

Ted J. Solomon, "Early Vaisnava Bhakti and its Autochthonous Heritage" *History of Religions,* Vol.10, No.1, Aug. 1970).

P.S. Srinivasachari, *Aspects of Advaita* (Mylapore: Sri Kirshna Library, 1949).

Claude Alan Stark, *God of All* (Cape Cod, Mass.: Claude Stark Inc., 1974).

Harold W. Turner, *The Idea of the Holy—A Guide for Students* (Aberdeen: Aberdeen People's Press, 1974).

Swami Vivekananda, The Complete Works of ... (Calcutta: Advaita Ashram, 1969).

Jacques Waardenburg, *Classical Approaches to the Study of Religion* (The Hague: Mouton, 1973).

Benjamin Walker, *The Hindu World,* 2 vols, (New York: Frederick A. Praeger, 1968).

A.K. Warder, *An Introduction to Indian Historiography* (Bombay: Popular Prakashan, 1972).

Webster's New International Dictionary (2nd edn), (Springfield Mass.: G. & S. Merriam & Co.).

Paramahamsa Yogananda, *Autobiography of a Yogi* (Los Angeles: Self - realization Fellowship, 1973).

R.C. Zaehner, *The Bhagavad Gita* (Oxford: Clarendon Press, 1965).

——, *Drugs, Mysticism and Make - believe* (London: Collins, 1972).

——, *Hindu and Muslim Mysticism* (London: Athlone Press, 1960).

——, *Hinduism* (Oxford University Press, 1962).

———' *Mysticism Sacred and Profane* (Oxford: Clarendon Press, 1957).

R.C. Zaehner, (ed.) *The Concise Encyclopaedia of Living Faiths* (Boston: Beacon Press, 1959).

INDEX